CAMBRIDGE MONOGRAPHS ON PHYSICS

GENERAL EDITORS

N. FEATHER, F.R.S.
Professor of Natural Philosophy in the University of Edinburgh

D. SHOENBERG, PH.D.
Fellow of Gonville and Caius College, Cambridge

OSCILLATIONS OF THE EARTH'S ATMOSPHERE

OSCILLATIONS
OF THE
EARTH'S ATMOSPHERE

BY

M. V. WILKES

Director of the University Mathematical Laboratory, Cambridge

CAMBRIDGE
AT THE UNIVERSITY PRESS
1949

PUBLISHED BY
THE SYNDICS OF THE CAMBRIDGE UNIVERSITY PRESS

London Office: Bentley House, N.W. 1
American Branch: 51 Madison Avenue, New York

Agents for Canada, India, and Pakistan: Macmillan

Printed in Great Britain at the University Press, Cambridge
(Brooke Crutchley, University Printer)

GENERAL PREFACE

The Cambridge Physical Tracts, out of which this series of Monographs has developed, were planned and originally published in a period when book production was a fairly rapid process. Unfortunately, that is no longer so, and to meet the new situation a change of title and a slight change of emphasis have been decided on. The major aim of the series will still be the presentation of the results of recent research, but individual volumes will be somewhat more substantial, and more comprehensive in scope, than were the volumes of the older series. This will be true, in many cases, of new editions of the Tracts, as these are republished in the expanded series, and it will be true in most cases of the Monographs which have been written since the War or are still to be written.

The aim will be that the series as a whole shall remain representative of the entire field of pure physics, but it will occasion no surprise if, during the next few years, the subject of nuclear physics claims a large share of attention. Only in this way can justice be done to the enormous advances in this field of research over the War years.

N. F.
D. S.

CONTENTS

AUTHOR'S PREFACE

I would like to express my gratitude to Professor S. Chapman, F.R.S., for his interest in this book and for allowing me to reproduce several illustrations from his Presidential Address to the International Association of Meteorology (Union of Geodesy and Geophysics). He was kind enough to lend me the original drawings and to inform me in advance of publication of several corrections.

I am also indebted to Mr K. Weekes for reading the manuscript and commenting on it.

M. V. WILKES

CHAPTER I

THE LUNAR AND SOLAR AIR-TIDES

1·1. Historical introduction

The invention of the barometer by Torricelli in 1643 was an event which attracted much attention. Other investigators soon constructed barometers and began to study the variations in the height of the mercury. Some of the variations were found to be due to defects in the instruments themselves, but even when these had been rectified it was still found that variations from day to day and from hour to hour took place, although two barometers at the same place always showed the same height.

The early experiments were conducted in temperate latitudes. By whom, and under what circumstances, a barometer was first taken to the tropics is not clear, but towards the end of the seventeenth century it was known in France that the height of the mercury was much more constant in the tropics than in higher latitudes, except during the passage of hurricanes. Beze, a French missionary, knew of this fact by hearsay, and verified it during a voyage to India in 1690. He did not, however, notice a phenomenon which later attracted the attention of the great eighteenth-century French mathematicians, namely, that there is a small daily variation in the height just large enough to be clearly discernible if the results for a few days are plotted. Fig. 1 shows such a record, taken in modern times, and it will be seen that the variation is principally semi-diurnal in character, with maxima at about 10 a.m. and 10 p.m. local solar time. A record from a non-tropical station is also shown for comparison.

During the eighteenth century, Newton's theory of universal gravitation was applied by Bernoulli, D'Alembert and others with outstanding success to many problems, including the theory of tides. It was realized that the gravitational forces which were the cause of tides in the oceans would also act on the atmosphere, and produce an air-tide. It was conjectured that the barometric variations observed in the tropics were of this nature. It had, however, to be explained why the air-tide apparently depended on the sun alone, whereas both sun and moon have their effect on the oceans, that of the moon preponderating in the ratio of approximately 11:5. Laplace answered this question by saying that the air-tide was mainly due to the thermal action of the sun, and that the purely gravitational effects of both moon and sun were of smaller magnitude.

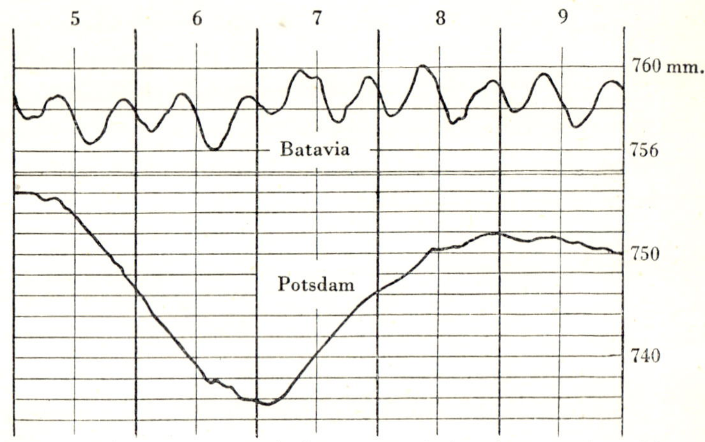

Fig. 1. A barometric record taken at a tropical station, together with one taken in a temperate latitude (Bartels).

During the next 50 years the solar barometric variation was studied at a number of stations in different parts of the world, and Kelvin in 1882 was able to quote a table showing the Fourier components with periods of 24, 12 and 8 hours for thirty different stations. He pointed out that, especially in high latitudes, the 12-hour component was larger than the 24-hour component, although the reverse would be expected on the thermal hypothesis. He suggested by way of explanation that the atmosphere had a free period which was nearer to 12 hours than to 24 hours. As this suggestion of Kelvin's is the origin of the 'resonance theory' with which this book is largely concerned, the actual passage in Kelvin's paper where it is made is worth quoting: 'The cause of the semi-diurnal variation of barometric pressure cannot be the gravitational tide-generating influence of the sun, because if it were there would be a much larger lunar influence of the same kind, while in reality the lunar barometric tide is insensible, or nearly so. It seems, therefore, certain that the semi-diurnal variation of the barometer is due to temperature. Now, the *diurnal* term, in the harmonic analysis of the variation of *temperature*, is undoubtedly much larger in all, or nearly all, places than the *semi-diurnal*. It is then very remarkable that the *semi-diurnal term of the barometric effect* of the variation of temperature should be greater, and so much greater as it is, than the diurnal. The explanation probably is to be found by considering the oscillations of the atmosphere, as a whole, in the light of the very formulas which Laplace gave in his *Mécanique céleste* for the ocean, and which he showed to be also applicable to the atmosphere. When thermal influence is substituted for gravitational, in the tide-generating force reckoned for, and when the modes of oscillation corresponding respectively to the diurnal and semi-diurnal terms of the thermal influence are investigated, it will probably be found that the period of free oscillation of the former agrees much less nearly with 24 hours than does that of the latter with 12 hours; and that, therefore, with comparatively small magnitude of the tide-generating force, the resulting tide is greater in the semi-diurnal term than in the diurnal.'

Laplace published his dynamical theory of the tides in 1778 and summarized it in the *Mécanique céleste*, the various volumes of which were published between 1799 and 1823. This work is the basis of a large part of modern tidal theory. Laplace was able to treat mathematically the problem of the oscillations of an ocean of uniform depth (and also certain cases of variable depth) on a rotating globe under the action of gravitational tide-producing forces. He showed further that the tidal oscillations of an atmosphere could be deduced from those of an ocean of depth equal to the scale height (i.e. height of the homogeneous atmosphere). The assumptions he made in deriving this result were: (1) that vertical acceleration could be neglected, (2) that the temperature of the atmosphere was uniform throughout, and (3) that the oscillations took place under isothermal conditions. Laplace was, of course, under the necessity of making simplifying assumptions in order to reduce the problem to a form in which it could be solved, but it is interesting to note that he appears to have been quite satisfied with the third of these assumptions, although he himself had corrected

Newton's value for the velocity of sound by pointing out that the changes of volume in a sound wave take place adiabatically. It is now realized that the same is true of atmospheric oscillations, which, although much longer in period, are on such a vast scale that flow of heat by conduction and convection is quite insufficient to bring about equalization of temperature during the oscillation.

As very little information was available concerning the thermal action of the sun, little progress could be made in treating the solar air-tide otherwise than qualitatively, but Laplace realized that if the lunar air-tide could be detected it would provide a means of testing his theory. His calculations led him to expect an amplitude of about $\frac{1}{4}$ mm. in the tropics, and he thought there would be a good chance of detecting this by averaging the data over a sufficiently long period. Unfortunately, no suitable data were available to him from a tropical station, so he resolved to try to determine the tide from measurements made at Paris. He was able to obtain for this purpose a series of readings taken four times a day over a period of 8 years. He decided to make no use of the readings taken at night since he considered that there might be a systematic error on account of the difficulty of reading accurately in a bad light, and he thus had nearly 5000 readings in all. From these he found by harmonic analysis a result which was only about a tenth as large as he had expected, and which he decided, after a detailed discussion based on the theory of errors, was not significant. If the tide were of this magnitude, he calculated that he would require no fewer than 40,000 observations to determine it accurately. This result was confirmed in 1828, when Bouvard made a recalculation using an additional 3 years' data, and got quite a different result. Later still, in 1843, Eisenlohr used 22 years' data (32,000 observations in all) and still failed to obtain a significant result. (See Chapman, 1939.)

Meanwhile the British Government had established, in tropical latitudes, a number of colonial observatories which included meteorological observations in their programme, and the credit for the first reliable determination of the lunar tide anywhere in the world goes to Lefroy, Director of the observatory at St Helena (1842) (Sabine, 1847). The fact that he was able to do this with only 17 months' observations, whereas Eisenlohr had failed to do so at Paris with 22 years', illustrates forcibly the greater prominence of the lunar atmospheric oscillation in the tropics as compared with temperature latitudes. Later (1847) Lefroy, with Smythe and Sabine, used 3 years' observations to obtain an improved determination. The amplitude was found to be about 0·06 mm.

Other tropical determinations followed. In 1852 Elliot published a result based on 5 years' observations at Singapore, and Bergsma in 1871 gave one based on 3 years' data obtained at Batavia.

Encouraged by these successes in searching for the lunar air-tide at tropical stations, various workers made renewed attempts to find it in higher latitudes. These attempts culminated in a very extensive investigation by Airy (1877) based on 180,000 hourly observations taken at Greenwich over a period of 20 years. His conclusion was that 'we can assert positively there is no trace of a lunar tide in the atmosphere'.

In 1918 Chapman approached the problem afresh, and at last succeeded in demonstrating the existence of the lunar tide at Greenwich. The reason for his success was that he included in the analysis only those days for which the daily barometric range was less than 0·1 in. In this way, although two-thirds of the available data were rejected, the random variations were so reduced that a significant result was obtained. The amplitude found for the semi-diurnal lunar tide was about 0·01 mm. of mercury, and the maxima occurred about $\frac{1}{2}$ hour before the moon's upper and lower transits. The amplitude of the tide is less than the error of reading the photographic barograph from which the data were obtained, and it is an interesting illustration of the theory of random errors that it should be possible to determine a systematic variation of this magnitude.

On the theoretical side Laplace's work was developed by Margules, who made various calculations based on the theory, and also discussed the mechanism of thermal action. The next big step, however, was taken by Lamb, who in 1910 showed that the oscillations of an atmosphere which was in adiabatic equilibrium, and in which the pressure changes were assumed to occur adiabatically, could also be discussed in terms of an equivalent ocean of uniform depth. It turned out that this depth was equal to the scale height of the atmosphere at its base, a result which was, rather surprisingly, the same as that obtained by Laplace for the isothermal oscillations of an isothermal atmosphere. As a result of this development the idea took firm root that, even if the atmosphere were not in adiabatic equilibrium, its oscillations could be discussed in terms of an equivalent ocean. Making use of a development of Laplace's theory due to Hough, Lamb showed that for the atmosphere to have a resonant mode of similar form to the observed semi-diurnal solar tide the depth of the equivalent ocean must be about 8 km. He also drew attention to the fact that if the dissipative forces were small the resonance would be quite sharp, so that in order to obtain appreciable amplification the resonant period would have to be within a few minutes of 12 solar hours. He pointed out that if this were the case the solar semi-diurnal oscillation would be greatly magnified as compared with the corresponding lunar oscillation, and suggested that the relative importance of the gravitational and thermal tide-producing forces might be reconsidered. This question must still be considered an open one.

By 1929 the evidence for the resonance hypothesis was accumulating, and it appeared that the theory was well on the way to being established. At about that time, however, G. I. Taylor showed that a value for the depth of the equivalent ocean could be obtained from a knowledge of the velocity of the propagation of very long air-waves in the atmosphere. Such waves had been observed on two occasions, the first being the Krakatoa eruption in 1883, and the second the fall of the Great Siberian Meteor in 1908. Measurements of the velocity with which these waves were propagated agreed closely, and gave an equivalent depth for the atmosphere of just over 10 km., a value appreciably greater than that demanded by the resonance theory.

This contradiction, which appeared fatal to the resonance theory, was resolved shortly afterwards (1936), when G. I. Taylor himself showed that the atmosphere might under certain circumstances have a whole series of equivalent depths, instead of one only as had been supposed up to that time. Pekeris, in an important paper published in 1937, examined the theory in detail, and showed that an atmosphere in which the temperature conformed to accepted ideas up to about 60 km. could be made to give two equivalent depths of the correct values, provided the temperature was assumed to fall to a low value again at higher level. The resonance theory was thus shown to be consistent with the measurements of the velocity of propagation of very long air-waves.

A striking consequence of Pekeris's theory is that at high level the pressure variation becomes reversed in phase and highly magnified. This removed a discrepancy between the theory of atmospheric oscillations and the dynamo theory of magnetic variations. Shortly afterwards, however, Appleton and Weekes (1939) detected and measured a lunar tide in the E region of the ionosphere (110 km.); interpreted in accordance with the usually accepted ideas on the formation of the E region, these measurements indicated that the pressure variation at this level was in phase with that at the ground, and not in anti-phase as Pekeris's theory would indicate. Quite recently the present writer, in collaboration with K. Weekes, has shown that this, and certain other difficulties raised by the same measurements, may be resolved if the temperature is supposed to rise after the fall required by Pekeris, to reach a further maximum in the E region. An alternative explanation has been given by Martyn.

1·2. The solar barometric variation

Extensive tabulations of the 24-, 12- and 8-hourly components of the barometric variation have been published by Hann (1889, 1892, 1918), and similar data, though not in such detail, have been given for the 6-hourly component by Pramanik (1926). It is not easy to make a selection from these data, but tables 1 and 2 have been drawn up to illustrate some of the main points. It must, however, be remembered that the conclusions summarized below are based on far more extensive data than are here given.

We denote the 24-, 12-, 8- and 6-hourly components by

$$s_1 \sin(t+S_1),\quad s_2 \sin(2t+S_2),\quad s_3 \sin(3t+S_3)\quad\text{and}\quad s_4 \sin(4t+S_4)$$

respectively, where t is local time. Table 1 shows the phase and amplitude for the same station—Montevideo—at different times of the year, while table 2 shows how these quantities vary from station to station during the same month—January. The figures

TABLE 1. *Montevideo* (10 *years' data*)

	s_1	S_1	s_2	S_2	s_3	S_3
Jan.	436	−24·5	367	144·5	110	142·0
Feb.	340	−28·7	403	144·9	89	123·1
Mar.	438	−23·3	400	145·6	35	72·5
Apr.	321	−11·0	404	153·0	79	20·1
May	344	− 1·3	377	153·6	134	4·7
June	292	− 1·0	385	153·3	143	−0·5
July	252	8·4	449	152·4	138	1·7
Aug.	288	7·0	444	154·6	114	−2·1
Sept.	331	− 2·4	468	160·5	54	39·4
Oct.	348	− 9·9	422	161·4	46	118·8
Nov.	428	−14·3	391	162·1	86	155·2
Dec.	365	−12·7	401	151·2	124	153·2
Year	341	−11·0	407	153·3	43	46·4

TABLE 2

Station	Latitude	s_1	S_1	s_2	S_2	s_3	S_3	s_4	S_4
Batavia	6·2S	537	17·0	998	156·9	10	203·8	32	60
Dar-es-Salaam	6·8S	787	−16·3	790	153·7	34	145·4	12	104
Bombay	18·9N	518	−27·6	1090	158·5	168	357·5	69	−120
Calcutta	22·5N	737	−29·6	1043	152·6	193	356·8	63	−135
São Paulo	23·6S	345	3·5	631	152·2	84	183·4	9	35
Curityba	25·4S	406	16·3	726	145·0	115	191·5	21	87
Johannesburg	26·2S	484	− 7·7	594	145·6	108	169·3	27	20
Kimberley	28·7S	808	− 2·3	605	155·2	91	180·9	—	—
Cordoba	31·4S	832	2·6	668	141·5	174	161·3	—	—
Rosario	32·1S	857	− 1·4	553	161·0	134	178·3	—	—
Montevideo	34·9S	436	−24·5	367	144·5	110	142·0	15	8
Melbourne	37·8S	338	12·3	577	161·4	124	175·1	—	—
Lisbon	38·7N	80	−21·2	461	154·5	182	353·2	—	—
Milan	45·5N	141	9·8	298	149·7	125	354·5	—	—
Geneva	46·2N	110	−19·5	350	166·5	90	3·8	—	—
Kremsmünster	48·1N	131	13·4	268	151·2	108	8·9	—	—
Munich	48·2N	86	−12·7	169	160·9	122	357·3	—	—
Prague	50·1N	178	3·2	196	140·8	106	354·1	—	—

for 6-hourly components in table 2 are taken from Pramanik, and refer to the summer or winter months according as the station is in the southern or northern hemisphere. Amplitudes are given in units of 0·001 mm. and phase angles in degrees.

The most striking feature of the data when viewed as a whole is the regularity of the 12-hourly component, which, except in polar regions, varies little (when expressed in terms of local time) either with longitude or with time of year. The amplitude, moreover, decreases uniformly with increase of latitude, and is, as Kelvin observed, on the whole slightly greater than that of the 24-hourly

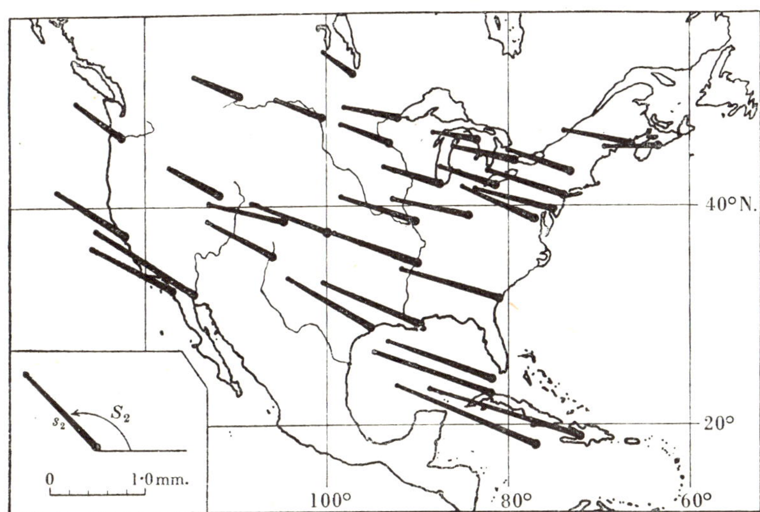

Fig. 2. Vectorial representation of $s_2 \sin(2t+S_2)$ (mean annual value) for stations in North America (Chapman, 1939).

component. The regularity of the 12-hourly component is well illustrated by fig. 2, taken from Chapman (1939), in which a vectorial representation of the mean annual magnitude and phase is given for a number of stations in North America. It will be seen that both magnitude and phase are little affected by proximity either to the coast or to the Rocky Mountains.

The 24-hourly component is by contrast very much more variable, both in the sense of having greater random fluctuations—so that a much longer period is required to evaluate it—and in the sense of depending on geographical factors such as height of station above sea-level. It also depends markedly on the season, being greatest in the summer, and is approximately in phase with the passage of the sun. Since the diurnal component of the solar tide-producing force, which arises from the inclination of the earth's axis to the plane of the ecliptic, cancels out when averaged over a year, the mean annual barometric variation must be due entirely to thermal action.

The 8- and 6-hourly components are both small in amplitude and very variable; both show marked seasonal or annual variations. The phase of the 8-hourly component becomes very erratic in the months of September and October, while that of the 6-hourly component is noticeably more regular in winter than in summer. Although the 8-hourly component is irregularly distributed over the earth's surface, there is a certain regularity in the annual variation, which affects stations in similar latitudes in the same way. Very roughly, the changes are equal and opposite in the two hemispheres. Schmidt (1919), working with Hann's data, showed that the amplitude of the 8-hourly component (if no attention is paid to the phase) can be represented approximately by the following empirical formulae:

$$(0{\cdot}02+0{\cdot}07\cos^2\theta)\sin^3\theta+0{\cdot}46\cos\theta\sin^3\theta \quad \text{(southern summer)},$$

$$(0{\cdot}02+0{\cdot}07\cos^2\theta)\sin^3\theta-0{\cdot}36\cos\theta\sin^3\theta \quad \text{(northern winter)},$$

the unit being 1 mm. and θ the co-latitude.

If the mean annual 12-hourly component is examined in more detail, it is found that near the poles the maxima and minima do not occur at the same local time as they do in low and medium latitudes, but instead tend to occur at the same Greenwich time. It was therefore suggested by Schmidt (1890) that the total oscillation should be regarded as made up of two components, a travelling wave which follows the sun round the earth, and a standing wave in which pressure is alternately high at the poles and at the equator.

Schmidt made certain calculations based on this idea using the data published by Hann, and gave the following tentative formulae in terms of spherical harmonic functions for the variation of the two components over the globe:

Travelling wave:

$$0{\cdot}906\sin^2\theta-0{\cdot}573\sin^2\theta\,(\cos^2\theta-\tfrac{1}{7}), \tag{1}$$

Standing wave:

$$-0{\cdot}042\cos\theta\sin(2t-2\phi+141{\cdot}8^\circ) +0{\cdot}216\,(\cos^2\theta-\tfrac{1}{3})\sin(2t-2\phi+128^\circ), \tag{2}$$

where t is local time, θ the co-latitude and ϕ the longitude. Schmidt's work, and that of Jaerish and Alt who followed him, was, however, not wholly satisfactory, in that they made certain arbitrary assumptions about the phase of the oscillations, and the matter was considered afresh by Simpson (1918) with the aid of data from 214 stations.

TABLE 3

Station	Latitude	Longitude	Time of max. local time (a.m. and p.m.)
San José	9° 56′ N.	84° 4′ W.	9·6
Quito	0° 14′ S.	78° 32′ W.	10·0
Para	1° 27′ S.	48° 29′ W.	10·0
Quixeramdbin	5° 16′ S.	39° 56′ W.	10·3
Ascension	7° 55′ S.	14° 25′ W.	9·8
Gabun	0° 25′ N.	9° 35′ E.	9·9
Cameroons	4° 3′ N.	9° 40′ E.	9·7
St Paul de Loanda	8° 49′ S.	13° 7′ E.	9·8
Kwai	4° 45′ S.	38° 18′ E.	10·0
Zanzibar	6° 10′ S.	39° 10′ E.	9·9
Dar-es-Salaam	6° 49′ S.	39° 19′ E.	9·6
Trivandrum	8° 31′ N.	76° 59′ E.	9·5
Singapore	1° 15′ N.	103° 51′ E.	9·8
Batavia	6° 11′ S.	106° 50′ E.	9·7
Finschhafen	6° 34′ S.	147° 50′ E.	9·5
Nauru	0° 26′ S.	166° 56′ E.	9·7
Jaluit	5° 55′ S.	169° 40′ E.	9·5

The different behaviour of the semi-diurnal variation at equatorial and polar stations is well illustrated in tables 3–6, which are taken from Simpson's paper. Table 3 gives the observed times of maximum pressure for a series of stations near to the equator, and it will be seen that with one exception the maxima occur between 9.30 and 10 local time. Similar information is given in table 4 for a number of stations near the north pole. Here the times bear no relation to local time, but all occur at approximately the same Greenwich time, ten out of fifteen being within the hour 10.30–11.30.

Simpson analysed the data by assuming that the semi-diurnal barometric variation at any point on the earth's surface could be represented by the sum of two sinusoidal components, one depending on local time and one on Greenwich time; i.e. he wrote

$$s_2 \sin(2t+S_2) = b \sin(2t+B) + c \sin(2t+C-2\phi),$$

where b, B, c, C depend on latitude but not on longitude, and t is local time. For the purpose of analysis he divided the data into groups according to latitude as shown in table 5. Within each

TABLE 4

Station	Latitude	Longitude	Time of maxima	
			Local time (a.m. and p.m.)	Greenwich time (a.m. and p.m.)
Point Barrow	71° 17′ N.	156° 40′ W.	0·7	11·1
Princess Royal	72° 48′ N.	117° 54′ W.	5·3	13·2
Mercy Bay	74° 6′ N.	117° 54′ W.	4·2	12·0
Griffith Island	74° 36′ N.	95° 18′ W.	7·1	13·5
Wellington Channel	75° 30′ N.	92° 10′ W.	6·4	12·5
Polarishaus	78° 18′ N.	72° 51′ W.	6·2	11·0
van Renselser Harbour	78° 36′ N.	70° 35′ W.	8·1	12·8
Fort Conger	81° 44′ N.	64° 45′ W.	7·2	11·5
Polaris Bay	81° 36′ N.	62° 15′ W.	7·4	11·5
Sabine Island	74° 32′ N.	18° 49′ W.	10·0	11·3
Jan Mayen	70° 59′ N.	8° 28′ W.	10·4	11·0
Cap Thorsden	78° 28′ N.	15° 42′ E.	11·7	10·6
Polhem	79° 50′ N.	16° 4′ E.	0·3	11·2
Nova Sembla	72° 23′ N.	52° 43′ E.	2·4	10·8
Ssagastir	73° 23′ N.	124° 5′ E.	7·5	11·2

TABLE 5

Zone	Mean latitude	No. of stations in zone
10° S. to 10° N.	0	17
10° N. to 25° N.	18	15
25° N. to 35° N.	30	12
35° N. to 45° N.	40	46
45° N. to 55° N.	50	60
55° N. to 65° N.	60	18
65° N. to 75° N.	70	14
75° N. to 85° N.	80	8

group the observations were reduced to sea-level, and corrected to the mean latitude. Values of b, B, c, C were determined by least squares so as to give the best fit with the data for each zone taken separately. In this way values for the parameters averaged over the zones were obtained. The results are given in table 6.

It will be seen that the phase of the travelling wave is very nearly the same for all zones, the greatest deviations from the mean being -5° and $+4^\circ$. The amplitude shows a smooth increase with latitude. This internal consistency is strong evidence that the form of

TABLE 6

Latitude	Equatorial vibration		Polar vibration	
	b (mm.)	B	c (mm.)	C
0	0·920	156° 50′	0·068	−3° 58′
18	0·835	155° 17′	0·082	−23° 13′
30	0·628	149° 7′	0·059	10° 26′
40	0·387	153° 56′	0·043	91° 4′
50	0·240	153° 1′	0·041	104° 27′
60	0·096	158° 3′	0·062	108° 23′
70	0·022 (70–80)	152° 53′ (70–80)	0·072	98° 34′
80			0·080	116° 27′

analysis applied has a physical significance. The amplitude is found to vary very nearly as $\sin^3\theta$, the best representation being given by

$$0{\cdot}937 \sin^3\theta \sin(2t+154^\circ). \qquad (3)$$

It is to be noted that no theoretical significance is attached to the factor $\sin^3\theta$, which is introduced from purely empirical considerations. The variation with latitude is not, in fact, as Simpson realized, in very good agreement with that calculated from tidal theory, and this aspect of the matter will be further discussed in Chapter V.

For the representation of the polar oscillation, Simpson took an expression of the form $K(\cos^3\theta-\frac{1}{3})\sin(2t-2\phi+\Phi)$, where ϕ is the longitude. This was suggested to him by theoretical considerations based apparently on Rayleigh's work (1890) in which the rotation of the earth was neglected. In this way he obtained the following formula for the polar component:

$$0{\cdot}137(\cos^2\theta-\tfrac{1}{3})\sin(2t-2\phi+105^\circ),$$

giving for the complete oscillation

$$s_2\sin(2t+S_2)=0{\cdot}937\sin^3\theta\sin(2t+154^\circ)$$
$$+0{\cdot}137(\cos^3\theta-\tfrac{1}{3})\sin(2t-2\phi+105^\circ). \qquad (4)$$

Simpson calculated the values of s_2 and S_2 given by this formula for all the stations used in his analysis, and showed that the average error was only a few degrees in phase and a few per cent in amplitude. It is to be noted that the analysis was based on stations in the northern hemisphere, or near the equator, since the greater part of the data available referred to these regions. Simpson found, however, that such results as were available for the southern hemisphere were, on the whole, in as good agreement with the formula as were the others. The data are, therefore, as far as they go, consistent with the hypothesis that the oscillation is symmetrical about the equator.

It is in the treatment of the polar oscillation that Simpson's work is open to criticism. Examination of the last column of table 4 shows that the expression $\cos^2\theta - \frac{1}{3}$ does not fit the results at all closely; in particular, the amplitude does not fall to zero when $\theta = \cos^{-1}(1/\sqrt{3})$, and the phase changes by only 110° from the pole to the equator instead of by 180°. This lack of agreement is illustrated in fig. 3. It cannot, therefore, be said that the observational data lend support to the choice of this particular expression as a representation of the polar oscillation. Moreover, the method of analysis loses its point when it is realized that rotation of the earth modifies the oscillation considerably, and that the amplitude is no longer correctly given by Rayleigh's theory. In any case, there is a whole series of oscillations having non-zero amplitude at the poles—given for a non-rotating earth by spherical harmonics of different orders—and it is not clear why one should be chosen rather than another.

This criticism is not intended to imply that Simpson's formula for the total oscillation is inaccurate; on the contrary, as shown by Simpson himself, it agrees closely with the data. This is because the polar component makes an appreciable contribution to the total oscillation only in the higher latitudes, where it is represented adequately by $0{\cdot}137(\cos^2\theta - \frac{1}{3})$. The formula is, however, misleading in that it suggests that this particular form of variation with latitude has been established.

A better procedure would be to fit a purely empirical formula to the data, as was done in the case of the travelling wave. The following formula, obtained in this way, is put forward as an alternative to the one given by Simpson:

$$(0{\cdot}07 - 0{\cdot}1|\cos\theta|)\sin 2(t-\phi) + 0{\cdot}075|\cos\theta|\cos 2(t-\phi).$$

Results calculated from this formula are included in fig. 3 for comparison with those given by Simpson's formula. The total oscillation is then

$$0{\cdot}937 \sin^3\theta \sin(2t+154^\circ)+(0{\cdot}07-0{\cdot}1|\cos\theta|)\sin 2(t-\phi) + 0{\cdot}075|\cos\theta|\cos 2(t-\phi).$$

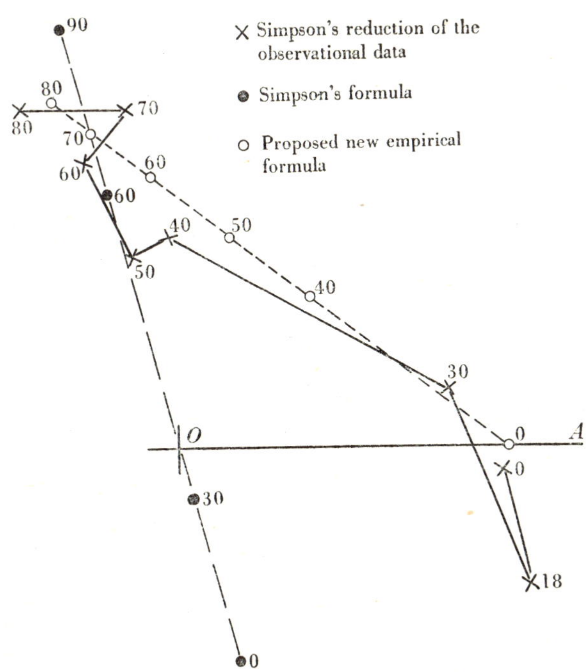

Fig. 3. Vector diagram showing phase and amplitude of the polar component of the solar semi-diurnal oscillation. The numbers against the points give the latitude in degrees.

The 12-hourly component exhibits a small but fairly regular annual variation. This may be seen in the results quoted in the table for Montevideo. The variation for Washington, D.C., is illustrated in fig. 4. A diagram of this kind in which the ends of the vectors representing the magnitude and phase of the oscillation are plotted is called by Bartels a 'harmonic dial'.

1·3. The lunar barometric variation

Chapman has followed up his remarkable achievement of isolating the lunar tide at Greenwich by evaluating it for a large number of other stations. For the more recent determinations Hollerith punched card machinery was used, a method of computation which showed great saving of labour compared with the hand methods used previously. Chapman summarized the present state of knowledge in a presidential address to the International Union

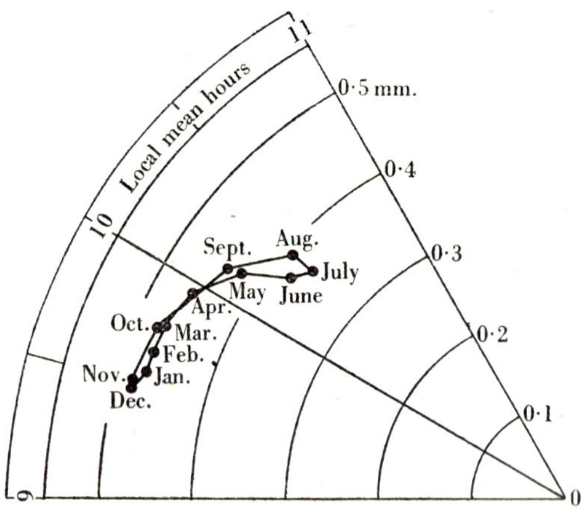

Fig. 4. Seasonal variation of $s_2 \sin(2t+S_2)$ for Washington, D.C. (Chapman, 1939).

of Geodesy and Geophysics (Association of Meteorology) in 1939. Much of the information given below is taken from this address. A later summary is given by Chapman and Tschu (1948).

The lunar tide is now known at fifty-four stations, including the three tropical stations previously mentioned. Of the new results forty-four are due to Chapman and his colleagues, and the remaining seven (three non-tropical) to Bartels, Pramanik, Robb and Tannahill.

As a result of these investigations it is possible to give a general account of the variations of the lunar air-tide over the earth's surface. It is found to be much less regular than the corresponding solar tide, and even now we have insufficient information to describe its variations completely.

In the first place the tide is almost entirely semi-diurnal, and may therefore be represented in the form $l_2 \sin(2\tau+L_2)$, where τ is the local lunar time. The diurnal component, which would be expected to exist except when the moon is near the equator, has not been found, the residual diurnal component obtained when the observations are reduced being no larger than would be expected to arise from statistical causes. The amplitude at the equator is about 0·06 mm. of mercury, and it decreases towards the poles in approximately the same way in each hemisphere. At Greenwich the amplitude is 0·0090 mm. The phase tends to lag slightly

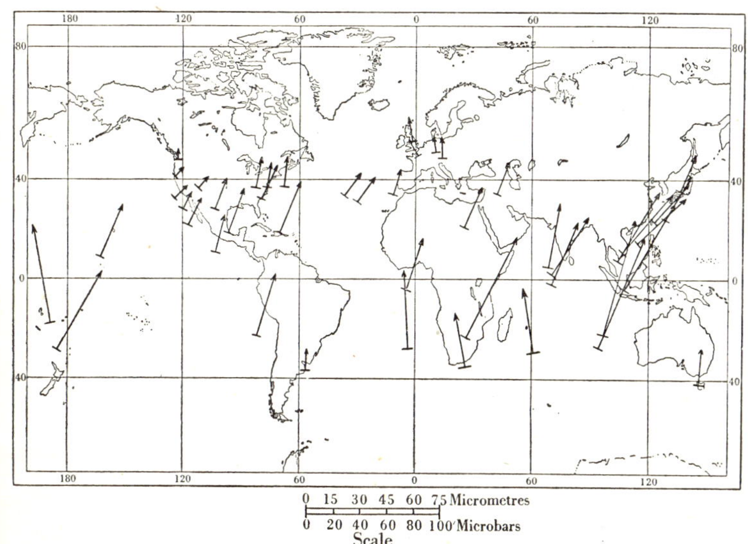

Fig. 5. Geographical distribution of amplitude and phase of annual lunar atmospheric tide (vector arrows *centred* at stations) (Chapman, 1939).

(about $\frac{1}{2}$ hour on the average) on the upper and lower lunar transits; there is some evidence that this lag is smaller in southern latitudes than elsewhere. At Greenwich the phase angle L_2 is 114°.

The data available are summarized in fig. 5, which shows, on an outline map of the world, the phase and amplitude of the semi-diurnal components of the lunar air-tide averaged over the whole year. In all later determinations Chapman has divided the data into groups corresponding to different seasons of the year, in order that variations of the phase and amplitude of the tide throughout the year can be followed. Dividing the data in this way into, say, n equal parts results in a lower accuracy for each component determination than for the mean annual tide determined from the data taken as a whole. This reduction is in the ratio $1:\sqrt{n}$, and it is found to be hardly practicable to divide the data into more than three parts corresponding to the following groups of months: May, June, July, August; November, December, January, February; March, April, September, October. The results of this more

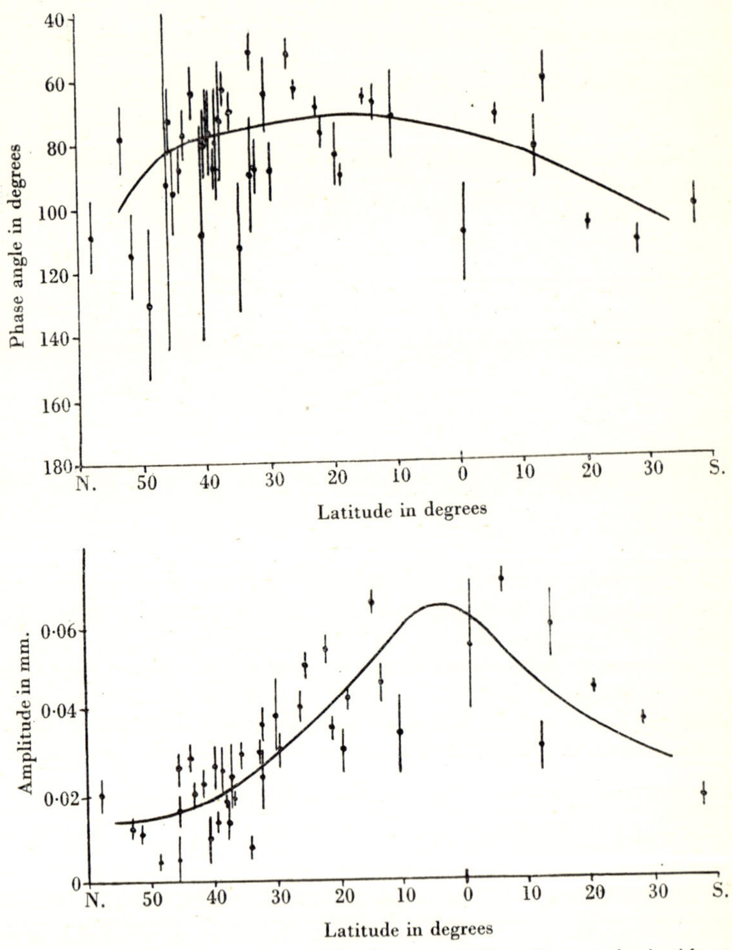

Fig. 6. Amplitude and phase of the lunar semi-diurnal atmospheric tide at about fifty stations (means for the four summer months).

detailed analysis are given in figs. 6–8. It will be seen that the phase of the lunar air-tide has an annual variation which takes the form of an extra retardation of phase, amounting to nearly an hour in the months grouped round December. This variation is an annual rather than a seasonal one, since it occurs in the northern winter and the southern summer. It was at one time thought that measurements of F_2 region ionization indicated a somewhat

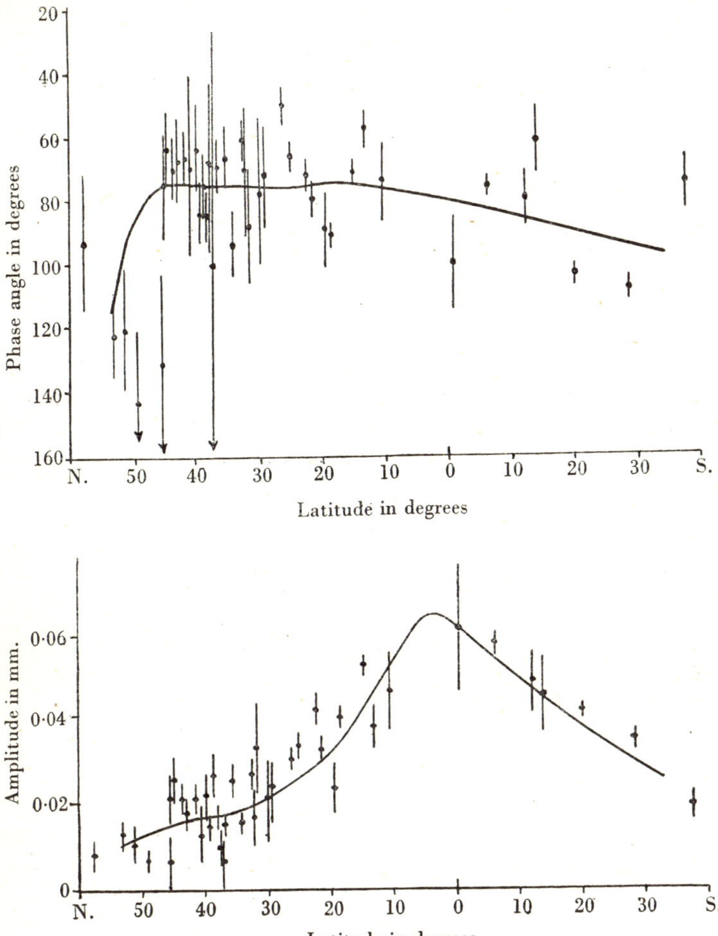

Fig. 7. Amplitude and phase of the lunar semi-diurnal atmospheric tide at about fifty stations (means for the four equinoctial months).

similar annual change in the ionosphere, but the greater quantity of data now available does not support this conclusion, especially when the asymmetry of the earth's magnetic field is taken into account.

It has been possible to obtain a more detailed analysis of the annual variation of the lunar air-tide by grouping together the observations from the three tropical stations Batavia, Bombay and

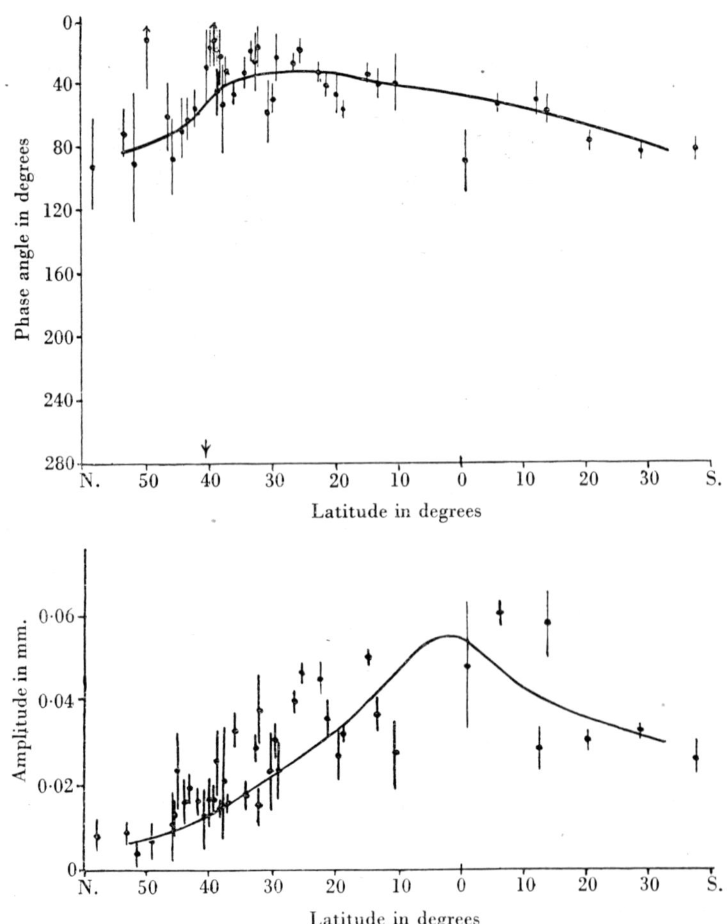

Fig. 8. Amplitude and phase of the lunar semi-diurnal atmospheric tide at about fifty stations (means for the four winter months).

Hongkong. This gives well over a million observations in all, and enables a determination to be made for each month of the year. The result is shown in fig. 10. The annual cycle is in an opposite direction to that of the solar semi-diurnal variation, and is of rather larger amplitude.

In addition to the general decrease of amplitude with increase of latitude, it appears that other factors, e.g. longitude and height of station, affect the tide. Stations near the west coast of North America give specially small amplitude, presumably on account

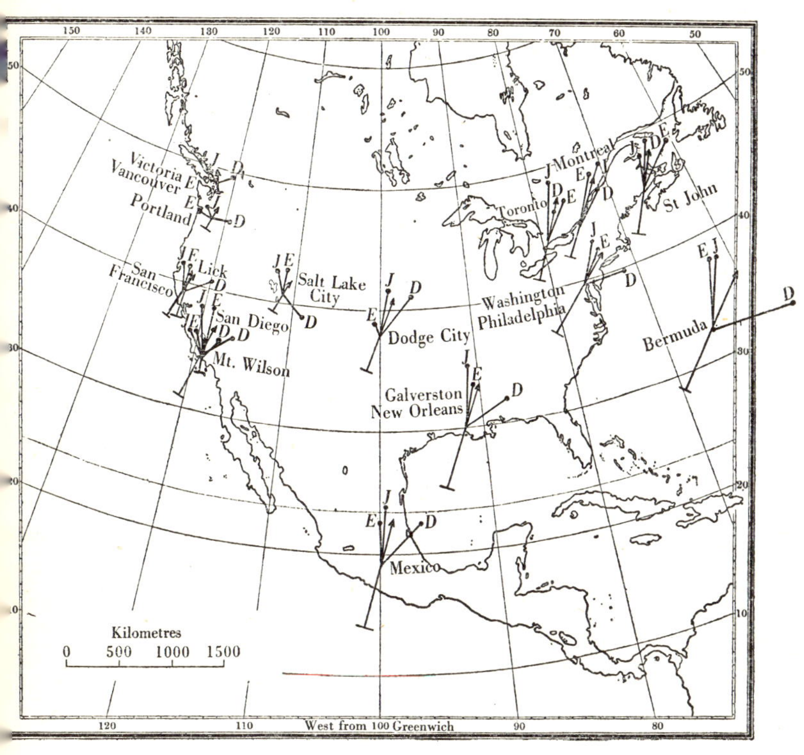

Fig. 9. Geographical distribution of amplitude and phase of lunar atmospheric tide; annual and 4-monthly means (vector arrows centred at stations) (Chapman, 1939).

of the proximity of the Rocky Mountains. This anomaly is well shown in fig. 9, which may be contrasted with a similar diagram (fig. 2) given previously for the solar oscillation.

If the lunar tide is produced solely by the gravitational action of the moon, its amplitude would be expected to vary inversely as the cube on the lunar distance. This effect was first looked for by

Lefroy in 1847 at St Helena. It is found that at some stations there is a change of the right order, but at others the results are not in good agreement with theory. It is hoped that further work will resolve this difficulty.

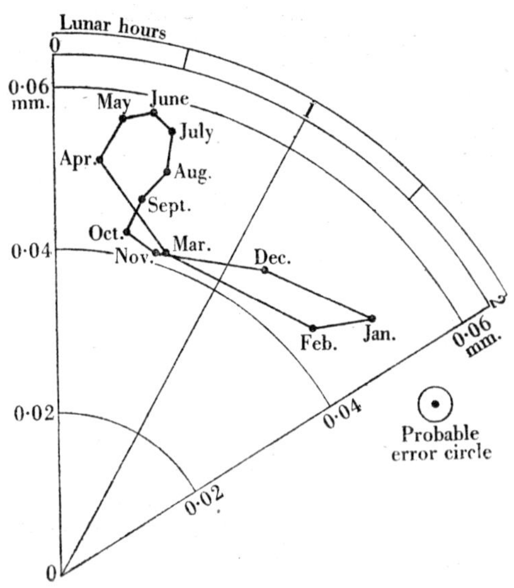

Fig. 10. Annual variation of the phase and amplitude of the lunar semi-diurnal tide obtained by averaging data from Batavia, Bombay and Hong-kong (Chapman, 1939).

It is well known that in ordinary sound waves the rarefaction and compression takes place under adiabatic conditions. Theoretical reasons will be given later for believing that, at any rate near the ground, the same is true of atmospheric oscillations, which, although much slower, are on such a large scale that heat transfer from one part of the atmosphere to another plays a negligible role. There should, therefore, be a semi-diurnal temperature change T associated with the lunar pressure change p given by

$$T = \frac{\gamma - 1}{\gamma} \frac{T_0}{p_0} p,$$

where T_0 and p_0 are the static temperature and pressure. Chapman has succeeded in checking this result directly by evaluating the lunar semi-diurnal temperature variation at Batavia. For this purpose he made use of 2-hourly observations for a period of no less than 62 years. The result is given in fig. 11, and will be seen to agree, within the limits of statistical expectation, with that calculated from the pressure variation.

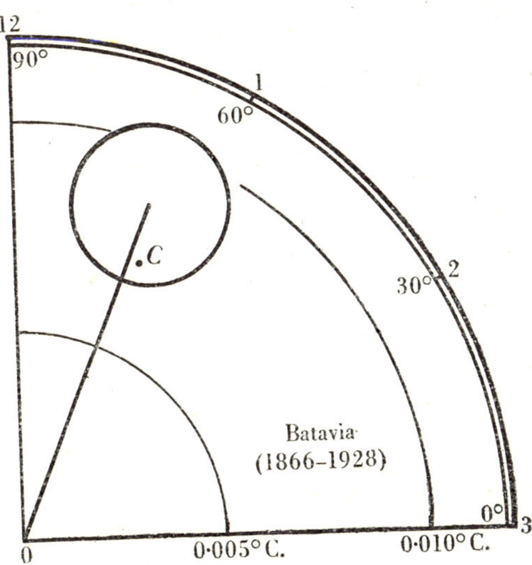

Fig. 11. Lunar semi-diurnal variation of air-temperature (point *C* indicates theoretical adiabatic variation) (Chapman, 1939).

1·4. Lunar tides in the ionosphere

The evaluation by Appleton and Weekes in 1939 of the lunar tide in the E region was the first satisfactory direct demonstration of the existence of a tidal effect at a level other than that of the ground. They used the well-known pulse technique in which a measurement is made of the time interval between the emission of a short pulse of electromagnetic energy and the return of an echo from the ionosphere. If half this interval is multiplied by the velocity of light *in vacuo*, the equivalent height of the region is obtained.

The degree of ionization required to reflect radio waves depends on their frequency, waves of high frequency penetrating more deeply into the region before being reflected than those of low frequency. The equivalent height is thus greater than the true height of the base of the region. Waves of frequency greater than a certain value—known as the critical penetration frequency—pass through the region altogether. When this frequency is approached the group velocity of the pulse is reduced and becomes less than that of light. The equivalent height is therefore greater than the height of the base of the region, even when allowance is made for penetration into the layer.

Experiment shows, however, that if the frequency is less than about three-quarters of the critical frequency the equivalent height is sensibly independent of frequency, although it has a diurnal variation depending on the zenith angle of the sun. Under these circumstances it may be assumed that the equivalent height differs little from the true height of the lower boundary of the region. For this reason Appleton and Weekes chose for their experiments a frequency of 1·8 Mcyc./sec., which is less than three-quarters of the critical frequency during nearly the whole of the day.

The equivalent height was measured to a nominal accuracy of 0·5 km. at every $\frac{1}{4}$ hr. of the day for a period of 12–14 days. The hourly means for the whole period were then plotted, and a smooth curve drawn through them. The deviations of the individual readings from this curve were tabulated, and the components of the lunar variation obtained by harmonic analysis. The use of this procedure enables the greater part of the solar variation to be eliminated before the harmonic analysis is begun.

It was found that the amplitude of the lunar diurnal variation was below significance level, but that there was a strongly marked semi-diurnal variation; it will be remembered from section 1·3 that the same was true of the lunar barometric variation. Fig. 12 shows in vectorial form the phase and amplitude of the semi-diurnal variation of E region height obtained from eleven different periods, each containing 12–14 days, covering the period 3 August 1937 to 11 July 1938. The points are closely grouped, a fact which is strong evidence that the result obtained is statistically significant. Further evidence on this point is provided by the point S which was calculated from a series of midday measurements made by the Radio Research Board at Slough during 1937. In this case the solar variation was eliminated by working in terms of the departures of individual daily readings from a 15-day running mean. It will be seen that the point S is not far removed from the cluster of points obtained by the other method.

Appleton and Weekes gave the following expression, based on

the whole of their data, for the lunar semi-diurnal variation of height of the base of the E region:

$$0 \cdot 93 \sin (2\tau + 112^\circ) \text{ km.},$$

where τ is the local lunar tide. The height at which this oscillation takes place is about 110 km. In view of the fact that the ionization in the E region is directly dependent on radiation from the sun, it

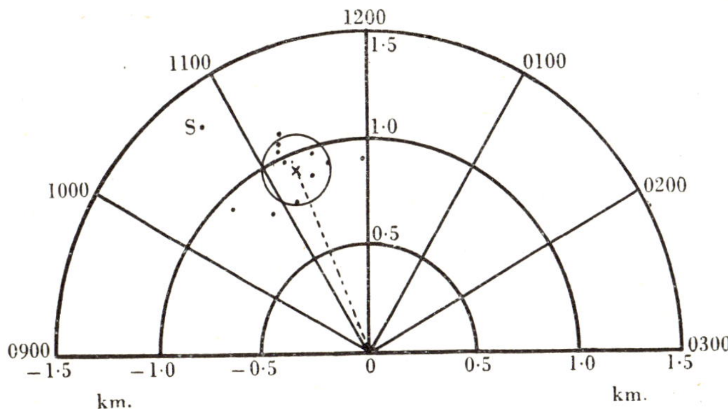

Fig. 12. Appleton and Weekes's determination of the lunar tide in the E region.

is not possible by radio means to determine the solar oscillation in the E region. The experiments under discussion show, however, that the ratio of the solar semi-diurnal oscillation to the lunar oscillation must be less than at the ground. If it were the same, the change in height of the E region due to the solar oscillation would be about 20 km.—the minimum height occurring at 4 p.m. In fact, the observed solar variation was 12 km., and the minimum height occurred at about midday. This is just about what would be expected from ionosphere theory without any tidal effect. Since the presence of a superposed solar oscillation of as much as 20 km. would modify the form of the variation to a very marked extent, we may conclude that the solar oscillation is in a smaller ratio to the lunar oscillation in the E region than it is at the ground.

CHAPTER II

THE THEORY OF OSCILLATIONS IN A ROTATING ATMOSPHERE

This chapter contains a theoretical discussion of the oscillations excited in a rotating atmosphere by gravitational action. The temperature of the atmosphere is assumed to vary with height in an arbitrary manner but to be independent of latitude and longitude. The present account follows closely that given by Pekeris (1937).

The next chapter deals with numerical evaluation of the theoretical tides. A physical interpretation of the mathematical theory will be given in Chapter IV, and readers who wish to do so may pass straight to this chapter.

2·1. General equations

The notation used is as follows:

a = radius of the earth,

ω = angular velocity of the earth,

g = acceleration of gravity,

γ = specific heat ratio,

k = Boltzmann's constant,

m = mean mass of air molecules,

θ = co-latitude,

ϕ = longitude,

z = height above the surface of the earth,

u = southward component of air velocity at (z, θ, ϕ),

v = eastward component of air velocity at (z, θ, ϕ),

w = vertically upward component of air velocity at (z, θ, ϕ),

$\chi(z, \theta, \phi)$ = divergence of velocity,

$$= \frac{1}{a \sin\theta}\frac{\partial}{\partial\theta}(u \sin\theta) + \frac{1}{a \sin\theta}\frac{\partial v}{\partial\phi} + \frac{\partial w}{\partial z}, \tag{5}$$

$\Omega(z, \theta, \phi)$ = tide-producing potential,

c = velocity of sound at height z,

$2\pi/\sigma$ = period of oscillation,

$f = \sigma/2\omega$.

We write p_0+p for the pressure at the point (z, θ, ϕ), p_0 being the static pressure, and p the variation due to the oscillation; similarly, $\rho_0+\rho$ is the density and T_0+T the temperature. p_0, ρ_0 and T_0 are functions of z but not of θ and ϕ. p, ρ, T, u, v and w are all small quantities whose squares and products may be neglected. The static pressure, density and temperature at the ground will be denoted by p_g, ρ_g and T_g. In the static case, i.e. when there is no oscillation, we have

$$\frac{\partial p_0}{\partial z} = -g\rho_0, \tag{6}$$

$$\frac{\partial p_0}{\partial \theta} = \frac{\partial p_0}{\partial \phi} = 0. \tag{7}$$

Since $p_0 = kT_0\rho_0/m$ it follows from (6) that

$$p_0 = p_g \exp\left(-\int \frac{mg}{kT_0}\, dz\right). \tag{8}$$

We have also the following expressions for the velocity of sound (Lamb, 1932, Chap. x):

$$c^2 = \gamma k T_0/m \tag{9}$$

$$= \gamma p_0/\rho_0. \tag{10}$$

We neglect the ellipticity of the earth, and the variation of the radius vector, gravity and $\partial\boldsymbol{\Omega}/\partial z$ with height. This approximation is most serious in the case of g, which is about 3% less at 100 km. than at the ground. Apart from these general assumptions we make the important assumption that the vertical acceleration is negligible. We have then for the equations of motion of the atmosphere

$$\frac{\partial u}{\partial t} - 2\omega v \cos\theta = -\frac{1}{a}\frac{\partial}{\partial \theta}\left(\frac{p}{\rho_0} + \boldsymbol{\Omega}\right), \tag{11}$$

$$\frac{\partial v}{\partial t} + 2\omega u \cos\theta = -\frac{1}{a\sin\theta}\frac{\partial}{\partial \phi}\left(\frac{p}{\rho_0} + \boldsymbol{\Omega}\right), \tag{12}$$

$$\frac{\partial p}{\partial z} = -g\rho - \rho_0\frac{\partial \boldsymbol{\Omega}}{\partial z}, \tag{13}$$

and for the equation of continuity

$$\frac{\partial \rho}{\partial t} + w\frac{\partial \rho_0}{\partial z} + \rho_0 \boldsymbol{\chi} = 0. \tag{14}$$

The adiabatic equation of state for the gas may be expressed in the form

$$\frac{Dp}{Dt}=c^2\frac{D\rho}{Dt}, \tag{15}$$

where D/Dt denotes 'differentiation following the fluid' (Lamb, 1932, p. 3). If the last equation is expanded, and use made of equations (6) and (14), we have

$$\frac{\partial p}{\partial t}=wg\rho_0-c^2\rho_0\chi. \tag{16}$$

The first step is to eliminate u, v, w, p and ρ from these equations in order to obtain a partial differential equation for $\chi\ (z, \theta, \phi)$. A series of special solutions will then be obtained by the method of separation of variables, and it will later be shown how these may be used to build up the general solution.

We take a time factor $e^{i\sigma t}$ and solve (11) and (12) for u and v, obtaining

$$u=\frac{\sigma}{4a\omega^2(f^2-\cos^2\theta)}\left[i\frac{\partial}{\partial\theta}+\frac{\cot\theta}{f}\frac{\partial}{\partial\phi}\right]\left(\frac{p}{\rho_0}+\Omega\right), \tag{17}$$

$$v=\frac{i\sigma}{4a\omega^2(f^2-\cos^2\theta)}\left[\frac{i\cos\theta}{f}+\frac{1}{\sin\theta}\frac{\partial}{\partial\phi}\right]\left(\frac{p}{\rho_0}+\Omega\right). \tag{18}$$

If these results are substituted in (5), we have

$$\chi-\frac{\partial w}{\partial z}=\frac{i\sigma}{4a^2\omega^2}F\left(\frac{p}{\rho_0}+\Omega\right), \tag{19}$$

where F denotes the differential operator

$$\frac{1}{\sin\theta}\frac{\partial}{\partial\theta}\left[\frac{\sin\theta}{f^2-\cos^2\theta}\left(\frac{\partial}{\partial\theta}-\frac{i\cot\theta}{f}\frac{\partial}{\partial\phi}\right)\right]$$
$$+\frac{1}{f^2-\cos^2\theta}\left[\frac{i\cot\theta}{f}\frac{\partial^2}{\partial\theta\partial\phi}+\frac{1}{\sin^2\theta}\frac{\partial}{\partial\phi^2}\right].$$

On differentiating equation (16) with respect to z we have, since

$$c^2\rho_0=\gamma p_0,$$

$$i\sigma\frac{\partial p}{\partial z}=wg\frac{\partial\rho_0}{\partial z}+\frac{\partial w}{\partial z}g\rho_0-c^2\rho_0\frac{\partial\chi}{\partial z}-\gamma\chi\frac{\partial p_0}{\partial z}.$$

If we substitute for $\partial p/\partial z$ from (13) and make use of (6), this becomes

$$-i\sigma\left(g\rho+\rho_0\frac{\partial\Omega}{\partial z}\right)=wg\frac{\partial\rho_0}{\partial z}+g\rho_0\frac{\partial w}{\partial z}-c^2\rho_0\frac{\partial\chi}{\partial z}+g\rho_0\gamma\chi.$$

We now use equation (14) to eliminate ρ, and obtain after re-arrangement

$$\frac{\partial w}{\partial z}=(1-\gamma)\chi+\frac{c^2}{g}\frac{\partial \chi}{\partial z}-\frac{i\sigma}{g}\frac{\partial \Omega}{\partial z}. \qquad (20)$$

We differentiate this equation and also equations (16) and (19) with respect to z, and eliminate $\partial w/\partial z$ and $\partial^2 w/\partial z^2$. This gives, on neglecting a term in $\partial^2\Omega/\partial z^2$,

$$c^2\frac{\partial^2\chi}{\partial z^2}+\left(\frac{dc^2}{dz}-\gamma g\right)\frac{\partial\chi}{\partial z}+\frac{1}{4a^2\omega^2}F\left\{\chi\left[g(1-\gamma)-g\frac{dc^2}{dz}\right]\right\}=0. \qquad (21)$$

To treat this equation by the method of separation of variables we put

$$\chi(z,\theta,\phi)=\chi(z)\,\psi(\theta,\phi)\,e^{i\sigma t}, \qquad (22)$$

and obtain the two equations

$$F[\psi(\theta,\phi)]+\frac{4a^2\omega^2}{gh}\psi(\theta,\phi)=0, \qquad (23)$$

$$c^2\frac{d^2\chi(z)}{dz^2}+\left(\frac{dc^2}{dz}-\gamma g\right)\frac{d\chi(z)}{dz}+\frac{\chi(z)}{h}\left[\frac{dc^2}{dz}+g(\gamma-1)\right]=0, \qquad (24)$$

where gh is a constant of separation of variables.

Equation (23) may be transformed into an equation in θ only by assuming that $\psi(\theta,\phi)$ varies with longitude like $e^{is\phi}$, i.e. by writing $\psi(\theta,\phi)=e^{is\phi}\Theta(\theta)$, where s is a constant. We then have

$$\frac{1}{\sin\theta}\frac{d}{d\theta}\left[\frac{\sin\theta}{f^2-\cos^2\theta}\frac{d\Theta(\theta)}{d\theta}+\frac{s\cot\theta}{f}\Theta(\theta)\right]$$
$$-\frac{1}{f^2-\cos^2\theta}\left[\frac{s\cot\theta}{f}\frac{d\Theta(\theta)}{d\theta}+\frac{s^2\Theta(\theta)}{\sin^2\theta}\right]+\frac{4a^2\omega^2}{gh}\Theta(\theta)=0. \qquad (25)$$

Equation (25) arises in connexion with the free oscillations of an ocean of uniform depth, and has been discussed by Laplace, Hough, and later writers on tidal theory. It will be considered in Chapter III.

2·2. Forced oscillations

We shall first suppose that forced oscillations of the atmosphere are under consideration, so that σ is given. The divergence $\chi(z,\theta,\phi)$ must be continuous at all points and such that the vertical velocity at the surface of the earth is zero. It is also necessary for $\chi(z,\theta,\phi)$ to satisfy a certain condition, the nature of which will be discussed later, at high level.

The condition of continuity clearly leads to the result that s must be an integer. It will be seen in the next chapter that, when applied to equation (25), it also leads to the result that if s and σ have prescribed values, h cannot be chosen arbitrarily, but is restricted to a series of discrete values. These we will denote by $h_r(s, \sigma)$ and the corresponding functions $\Theta(\theta)$ by $\Theta_r^s(\theta)$; the mode of oscillation will be referred to as the (s, r) mode. Although it is not shown explicitly by the present notation it must be remembered that the form of $\Theta_r^s(\theta)$ depends on h (and therefore on the period) as well as on s and r.

The assumption is now made that $\boldsymbol{\Omega}$ has the same variation over the surface of the earth as has $\boldsymbol{\chi}$, i.e. that

$$\boldsymbol{\Omega}(z, \theta, \phi) = \Omega(z)\, \Theta_r^s(\theta)\, e^{i(\sigma t + s\phi)}. \tag{26}$$

It is then possible to find a solution in which the variables u, v, w and p also vary in this way with θ and ϕ. Making this assumption we obtain from (19), (20) and (23)

$$\frac{p}{\rho_0} = \left\{ -\Omega(z) + \frac{h}{i\sigma}\left[-\gamma g \chi(z) + c^2 \frac{d\chi(z)}{dz} \right] \right\} \Theta_r^s(\theta)\, e^{i(\sigma t + s\phi)}, \tag{27}$$

where a term $hd\Omega/dz$ which is of order $h\Omega/a$ has been neglected in comparison with Ω. From (16) we now deduce that

$$w = \left\{ \frac{c^2}{g}\chi(z) + h\left[-\gamma\chi(z) + \frac{c^2}{g}\frac{d\chi(z)}{dz} \right] - \frac{i\sigma}{g}\Omega(z) \right\} \Theta_r^s(\theta)\, e^{i(\sigma t + s\phi)}. \tag{28}$$

Equations (11) and (12) then give u and v. It may be verified by direct substitution that these expressions for u, v, w and p satisfy the original differential equations.

Equation (24) for $\chi(z)$ may be simplified by writing

$$c^2 = \gamma g H, \tag{29}$$

$$x = \int \frac{dz}{H}, \tag{30}$$

$$\chi(z) = e^{\frac{1}{2}x} y(x). \tag{31}$$

H is the scale height of the atmosphere at the level under consideration†; in terms of temperature, $H = kT_0/mg$, where m is the mean molecular mass. Equation (24) becomes

$$\frac{d^2y}{dx^2} + \left[-\frac{1}{4} + \frac{1}{h}\left(\frac{dH}{dx} + \frac{\gamma - 1}{\gamma} H \right) \right] y = 0. \tag{32}$$

† Taylor and Pekeris used the symbols H and h with their meanings reversed. The notation given has been adopted here, since the use of H for scale height has now become fairly general.

Expressions for the velocities and the pressure variation in terms of y are given below:

$$u = \frac{\gamma g h\, e^{\frac{1}{2}x}}{4a\omega^2(f^2 - \cos^2\theta)}\left(\frac{dy}{dx} - \tfrac{1}{2}y\right)\left(\frac{d}{d\theta} - \frac{s\cot\theta}{f}\right)\Theta_r^s(\theta)\, e^{i(\sigma t + s\phi)}, \quad (33)$$

$$v = \frac{i\gamma g h\, e^{\frac{1}{2}x}}{4a\omega^2(f^2 - \cos^2\theta)}\left(\frac{dy}{dx} - \tfrac{1}{2}y\right)\left(\frac{\cos\theta}{f}\frac{d}{d\theta} - \frac{s}{\sin\theta}\right)\Theta_r^s(\theta)\, e^{i(\sigma t + s\phi)}, \quad (34)$$

$$w = \left\{-\frac{i\sigma}{g}\Omega + \gamma h\, e^{\frac{1}{2}x}\left[\left(\frac{H}{h} - \frac{1}{2}\right)y + \frac{dy}{dx}\right]\right\}\Theta_r^s(\theta)\, e^{i(\sigma t + s\phi)}, \quad (35)$$

$$p = \frac{p_g}{H}\left[-\frac{\Omega\, e^{-x}}{g} + \frac{\gamma h}{i\sigma}e^{-\frac{1}{2}x}\left(\frac{dy}{dx} - \tfrac{1}{2}y\right)\right]\Theta_r^s(\theta)\, e^{i(\sigma t + s\phi)}. \quad (36)$$

When expressed in terms of x, equation (8) becomes

$$p_0 = p_g\, e^{-x}. \quad (37)$$

Solutions of the type given here were first obtained by Taylor (1936) for the case of free oscillations ($\Omega = 0$). Taylor noted that the possibility of obtaining solutions in the ordinary form for the free oscillations of an incompressible ocean depended on the fact that $p/\rho_0(\chi - \partial w/\partial z)$ was constant throughout the ocean. He found that if he made the same assumption in the case of a compressible atmosphere he was led to differential equations similar to (24) and (25) with $\Omega = 0$. It may easily be verified from equations (19), (23) and (27) that in the solution presented here (with Ω put equal to zero) $p/\rho_0(\chi - \partial \omega/\partial z)$ does not depend on z, θ or ϕ.

2·3. Boundary conditions

At the surface of the earth the vertical component, w, of the velocity must vanish, i.e.

$$\left[\frac{dy}{dx} + \left(\frac{H}{h} - \frac{1}{2}\right)y\right]_{x=0} - \frac{i\sigma\Omega(0)}{\gamma g h} = 0. \quad (38)$$

To find the nature of the boundary condition to be imposed at high level we must consider the rate of flow of energy in a vertical column of air of constant cross-section. Examination of the left-hand side of equation (36) shows that, provided $dy/dx - \frac{1}{2}y$ does not decrease rapidly with height (an assumption which may be verified later), the term in Ω may be neglected when x is large. At high level, therefore, u is in quadrature with p, and there is on the average no flow of energy towards the poles. On the other hand, v is in phase with p, so that there is in general a non-zero flow of energy along the meridians of magnitude (when averaged over a cycle)

$$\frac{\gamma^2 g h^2}{8a\omega^2}\frac{p_g}{H\sigma}\left|\frac{dy}{dx}-\tfrac{1}{2}y\right|^2\frac{\Theta_r^s(\theta)}{f^2-\cos^2\theta}\left(\frac{\cos\theta}{f}\frac{d\Theta_r^s(\theta)}{d\theta}-\frac{s\Theta_r^s(\theta)}{\sin\theta}\right).$$

This expression is independent of ϕ, so that the amount of energy entering the column of air horizontally from an easterly direction is equal to the amount leaving in a westerly direction. This is true at all heights, so that in studying the balance of energy in the column we need pay no further attention to horizontal flow.

To find the rate at which energy is flowing in a vertical direction, we need to multiply together the real parts of the expressions for w and p given in equations (35) and (36), and then take the mean value. It is easily shown that this is equivalent to forming the expression

$$W=\tfrac{1}{2}\mathscr{R}pw^*,$$

where we use the notation $\mathscr{R}$ (or $\mathscr{I}$) to denote that the real (or imaginary) part is to be taken and * to indicate the conjugate complex. This gives, if it is assumed that the level is high enough for the term in Ω to be neglected,

$$W=\tfrac{1}{2}(p_g h\gamma^2/\sigma)[\Theta_r^s(\theta)]^2\mathscr{I}[y^*(dy/dx)]. \tag{39}$$

We can now consider the nature of the boundary conditions to be satisfied by $y(x)$ at high level. It is convenient first to assume that the scale height is independent of height; equation (32) then becomes

$$\frac{d^2y}{dx^2}+\left(\frac{H}{h}\frac{\gamma-1}{\gamma}-\frac{1}{4}\right)y=0. \tag{40}$$

We must consider two cases:

(*a*) when $$\left(\frac{H}{h}\frac{\gamma-1}{\gamma}-\frac{1}{4}\right)=-\lambda^2\text{ (say)}<0,$$

(*b*) when $$\left(\frac{H}{h}\frac{\gamma-1}{\gamma}-\frac{1}{4}\right)=\mu^2\text{ (say)}>0.$$

In case (*a*), the two solutions of equation (40) are

$$y=A\,e^{\pm\lambda x}, \tag{41}$$

and we choose the one with the lower sign. $y(x)$ is then bounded as x tends to infinity, and substitution in equation (39) shows that the rate of flow of energy across a horizontal plane tends to zero. In case (*b*), the two solutions of equation (40) are

$$y = A\,e^{\pm i\mu x}, \tag{42}$$

so that $y(x)$ remains finite in both cases as x tends to infinity. In order to choose between these two solutions, we note that the energy which is being supplied to the system through the action of the tide-producing force is mostly introduced at low level where the air density is greatest. At a sufficiently high level, therefore, the flow of energy must be in an upward direction. Use of equation (39) shows that the solution which satisfies this condition is obtained by taking the upper sign in equation (42).

The same consideration, namely, that the vertical flow of energy at high level, if not zero, must be upwards, serves to determine the appropriate solution when H is not independent of height.

An alternative treatment of the boundary condition at high level is to consider flow of energy out of a sphere concentric with the earth and of radius $a+z$. The amount of energy leaving the sphere in unit time is (if we neglect a term which tends to zero at high level)

$$\frac{1}{2}\left[\frac{p_g h\gamma^2}{\sigma}\int[\Theta_r^s(\theta)]^2 dS\right]\mathscr{I} y^* \frac{dy}{dx},$$

the integration being performed over the surface of the sphere. Since the quantity in the square brackets is positive, this leads as before to the conclusion that $\mathscr{I} y^*(dy/dx)$ must be positive. The advantage of this method of approach is that any consideration of horizontal flow of energy is avoided.

2·4. General solution

The oscillation considered above corresponds to the special form of tide-producing potential given by (26). It is easy to see on physical grounds that the general solution is to be obtained by superposing, in the right ratios, a series of solutions of the same type, but corresponding to different values of s and r. This may be proved mathematically by noting that the functions $\Theta_r^s(\theta)$ and $e^{is\phi}$ both form complete sets.

The above argument is equivalent to saying that the atmosphere has an infinite number of modes of oscillation, which are excited to varying degrees by the applied force. In order to facilitate the discussion it will generally be assumed that only one mode is excited at a time. This will often be approximately the case in practice, especially when the period under consideration is not far removed from a resonant period, but the possibility of two or more modes being excited to an approximately equal degree must not be overlooked.

The predominating components of $\mathbf{\Omega}$ have periods of half a lunar day and half a solar day respectively. A discussion of the form of $\mathbf{\Omega}$ is given by Lamb (1932, Chap. VIII, Appendix).

2·5. Free periods

From the equations given in the last section it is possible to determine the pressure variation at a point on the earth's surface produced by an exciting force of given period exciting a specified mode of oscillation. It may be found that when the period has certain values the amplitude becomes infinite; these are the free periods.

An alternative method of proceeding, if only the free periods are required, is to attempt to solve equation (32) subject to the additional condition that the pressure variation vanishes at the ground in the absence of an exciting force. This condition, which differs from (38) in that it is homogeneous in y, may be written

$$\left[(H-\tfrac{1}{2}h)\,y+h\frac{dy}{dx}\right]_{x=0}=0. \tag{43}$$

If free periods exist, it will be found that solutions satisfying this condition can be found, provided that h has one of a series of eigenvalues. These we shall denote by $h_n^{(f)}$, where the superscript denotes that free periods are involved, and where $n=1, 2, 3, \ldots$. Since, however, we already know from section 2·2 that h can only have the discrete values $h_r(s, \sigma)$ we must have $h_n^{(f)}=h_r(s, \sigma)$. Regarded as an equation for σ, this is the period equation for free oscillations. Corresponding to each $h_n^{(f)}$ there is a doubly infinite set of roots, and hence a doubly infinite set of free periods.

2·6. Propagation of a pulse

We shall now show how the general equations of section 2·2 may be used to discuss the propagation in the atmosphere of a pulse whose dimensions are small compared with the radius of the earth. For this purpose we return to equation (25), and resolve the pulse into its Fourier components. The disturbances to which we shall later wish to apply this theory are of the order of a few hundred km. in length and travel with a velocity of about 300 m./sec., so that we may assume that the main contribution comes from components which have a period small compared with that of rotation of the earth, i.e. we may make $f=\sigma/2\omega$ large. Equation (25) then becomes

$$\frac{1}{\sin\theta}\frac{\partial}{\partial\theta}\left(\sin\theta\frac{\partial\psi(\theta,\phi)}{\partial\theta}\right)+\frac{1}{\sin^2\theta}\frac{\partial^2\psi(\theta,\phi)}{\partial\phi^2}+\frac{a^2\sigma^2}{gh}\psi(\theta,\phi)=0. \quad (44)$$

If we confine attention to an area of the earth's surface sufficiently small to be regarded as flat, and take x and y to be Cartesian co-ordinates measured in a northerly and easterly direction respectively, we may write

$$dx=ad\theta, \quad dy=a\sin\theta d\phi.$$

As a further consequence of the small size of the pulse, the variation with θ of $(\partial/\partial\theta)\psi(\theta,\phi)$ is much more rapid than that of $\sin\theta$, so that (44) becomes

$$\frac{\partial^2\psi}{\partial x^2}+\frac{\partial^2\psi}{\partial y^2}+\frac{\sigma^2}{gh}\psi=0, \quad (45)$$

where ψ is now regarded as a function of x and y. This shows that simple harmonic waves are propagated with phase velocity $\sqrt{(gh)}$. Since this velocity is independent of σ it follows that a pulse of small dimensions is propagated, without change of form, with the same velocity.

It will be observed that (45) does not contain either the angular velocity ω, or the radius a of the earth. The same equation can in fact be obtained more simply by assuming at the outset that the earth is flat and non-rotating. If this is done, it will be found that the approximation of neglecting the vertical acceleration, which we made at the beginning, is equivalent to assuming that the waves are long compared with h. For this reason it is usual to refer to air-waves of the present type as long waves.

2·7. Atmosphere in adiabatic equilibrium

Determination of the quantities $h_n^{(f)}$ involves a knowledge of the variation of H with height, and is, in general, difficult. For an atmosphere in adiabatic equilibrium, however, it becomes quite simple. In this case it may easily be shown that

$$\frac{dH}{dz} = -\frac{\gamma - 1}{\gamma},$$

or in terms of x

$$\frac{dH}{dx} = -\frac{\gamma - 1}{\gamma} H,$$

so that equation (32) reduces to

$$\frac{d^2y}{dx^2} - \tfrac{1}{4}y = 0.$$

This equation has two linearly independent solutions, of which the one which is bounded as x tends to infinity is $y = A\,e^{-\frac{1}{2}x}$. The condition at the ground (equation (43)) will be satisfied provided that

$$h = H_g$$

where H_g is the scale height at the ground. Thus there is in this case just one $h_r^{(f)}$, equal to the scale height of the atmosphere at the ground.

This result was first obtained by Lamb, although it had been proved much earlier by Laplace for the case of isothermal oscillations in an atmosphere in isothermal equilibrium. This may be regarded as a special case in which γ is put equal to unity.

2·8. Damping

Degradation of the ordered motion of oscillation into random motion takes place partly through the action of viscous forces, and partly as a result of the transfer of heat by conduction or radiation from parts of the atmosphere in a state of adiabatic compression to parts in a state of adiabatic rarefaction. These processes may be subdivided as follows:

(1) *Molecular conductivity* (transport of heat by molecular diffusion).

(2) *Eddy conductivity* (transport of heat by eddy motion).

(3) *Molecular viscosity* (transport of momentum by molecular diffusion).

(4) *Eddy viscosity* (transport of momentum by eddy motion).

(5) *Radiation.*

The simplest of these processes to treat theoretically is molecular conductivity. To do this it is only necessary to replace the adiabatic equation of state (equation (15)) in the treatment given in section 2·1 of this chapter by the equation of conduction of heat which, for periodic motion, takes the form (cf. Lamb, 1932, p. 649)

$$K\frac{\partial^2 T}{\partial z^2}=i\sigma T+(\gamma-1)T_0\chi, \tag{46}$$

where K is the diffusivity. If K is put equal to zero, it may be shown that the two equations are equivalent.

We now require to find the circumstances in which it is justifiable to neglect thermal conductivity. To do this we substitute the solution obtained on this assumption in the neglected term in equation (46), and express the condition that the result is numerically much smaller than one of the terms retained, i.e. that

$$\left|K\frac{d^2T}{dz^2}\right|\ll|\sigma T|. \tag{47}$$

It is sufficient to assume that the atmosphere is isothermal at the level considered. If K is neglected, the solution is that given in section (a), and we easily obtain from equation (32)

$$y=\mathrm{e}^{\pm i\mu x} \quad \text{or} \quad \chi=\mathrm{e}^{(\frac{1}{2}\pm i\mu)x},$$

where

$$\mu^2=-\frac{1}{4}+\frac{\gamma-1}{\gamma}\frac{H}{h}.$$

Equation (46) then gives, K being neglected,

$$T=i\frac{\gamma-1}{\sigma}T_0\,\mathrm{e}^{(\frac{1}{2}\pm i\mu)x}.$$

Substituting this value in (47), we have

$$K\ll\frac{\sigma H^2}{\mu^2+\frac{1}{4}}. \tag{48}$$

If we take for purposes of illustration $H=6$ km., $h=7$ km. and $\sigma=\pi/21{,}600$ (corresponding to solar semi-diurnal oscillations), this inequality becomes

$$K\ll 2{\cdot}1\times 10^8. \tag{49}$$

K is proportional to the mean free path of the gas molecules, and therefore inversely proportional to pressure. Near the ground it is approximately equal to 0·26 and (49) is amply satisfied.

In order to get an idea of the level at which (49) ceases to be satisfied, we may take the atmosphere specified in the table at the end of Chapter IV. We find that at 120 km. $K=10^7$. Damping due to molecular conductivity thus first becomes of importance at about this level.

It can be shown that to a sufficient degree of approximation conduction of heat by eddy motion may be treated in a similar way to molecular conductivity, except that the eddy diffusivity K' must be used in equation (48) instead of the molecular diffusivity K (Brunt, 1939, p. 226). Near the ground K' is enormously greater than K, its value varying considerably according to the state of stability of the atmosphere. For example, from measurements made on the Eiffel Tower, Taylor obtained values of K' of the order of 10^5, while in an inversion over the Great Banks of Newfoundland he obtained values of the order of 10^3. Both these values are sufficiently small to satisfy with a large margin the inequality for K' analogous to (48).

No experimental information is available covering the order of magnitude of K' in the high atmosphere, and it might be expected to vary considerably at different levels. There is, however, no reason to believe that it would differ by any large factor from its value at the ground, and it would appear safe to assume that the effect of eddy conductivity is negligible at all levels.

The effect of molecular viscosity may be discussed by introducing appropriate terms into the equations of motion. The discussion follows the lines of that for molecular conductivity, and an inequality similar to (48) is obtained for the coefficient of molecular viscosity. The two effects are therefore of the same order of magnitude, and become of importance at about the same level in the atmosphere. This is as would be expected in view of the close physical connexion between the two phenomena.

Discussion of eddy viscosity is more complicated than the corresponding case of eddy conductivity, and no satisfactory theory exists at the present time. The close relationship already noted as existing between thermal conductivity and viscosity, however, provides strong grounds for supposing that the effect of eddy viscosity is of a similar order of magnitude to that of eddy conductivity at all levels.

Estimates of the rate of heat transfer by radiation in the very low atmosphere suggest that its effect in damping oscillations is intermediate in magnitude between those of molecular and eddy thermal conductivity. Any estimate of its magnitude at higher levels is difficult to make, since it will depend markedly on the distribution assumed for such gases (mainly water vapour and ozone) as have strong absorption bands in the infra-red. All that can be done is to make the tentative assumption that the rate of absorption due to heat exchange by radiation does not anywhere exceed its value at the ground by more than, say, 10^4, in which case it will be negligible.

It will be seen that if the arguments set out above are correct the effects of the various damping processes are negligible below about 120 km., and that above this level the rapid increase of the mean free path of the gas molecules causes molecular transport effects to play a predominating role. Some evidence that this is correct is provided from the study of atmospheric oscillations, since as will become clear in Chapter IV the existence of any appreciable damping below say 90 km. would be fatal to the resonance theory.

The discussion of the boundary condition at high level given in section 2·3 may now be supplemented by saying that the energy flowing upwards is ultimately absorbed as a result of the effects of viscosity and thermal conduction. Since the onset of these two effects with height is gradual, the absorption takes place without any appreciable reflexion of energy. The direction of flow just below the level at which absorption begins must therefore be upward. The boundary condition to be applied is thus in effect the same as in the idealized case considered previously, when absorption was neglected. The phenomenon is analogous to the absorption of electromagnetic energy in a wave-guide in which a thin wedge of absorbing material has been placed.

CHAPTER III

NUMERICAL EVALUATION OF AIR-TIDES

3·1. Laplace's tidal equation

The history of equation (25), which will be referred to as Laplace's tidal equation, is of some interest. It was first obtained by Laplace in the course of his investigation of the free oscillations of an ocean of uniform depth on a rotating earth; he later showed that it was also relevant to the free oscillation of an isothermal compressible atmosphere. It now arises in connexion with the forced oscillations of an atmosphere with an arbitrary temperature distribution.

Laplace first attempted a treatment of the equation by means of spherical harmonics, but found that this method presented difficulties. He therefore used instead a method in which a solution for Θ was obtained as a power series in $\sin\theta$. The series did not converge very rapidly, but Laplace was able to obtain a number of numerical results.

Some 120 years later, another and successful attempt to apply spherical harmonics to the problem was made by Hough. Although the series so obtained are more rapidly convergent than those in $\sin\theta$, numerical evaluation of the results still requires considerable labour, most of which is spent in evaluating continued fractions.

One advantage of the method from the point of view of tidal theory is that it is a simple matter to allow for the mutual attraction of the particles of the water comprising the ocean, an effect which Laplace had been compelled to neglect. This means, however, that the extensive calculations made by Hough are not as they stand directly applicable to the atmosphere, since the corresponding effect is there quite negligible. Fortunately, however, in the cases of most interest the necessary corrections are not difficult to make, and further reference will be made to this point later.

Hough showed that solutions for Θ representing free oscillations of the ocean could be obtained in the form of a series of associated Legendre functions, $P_n^s(\cos\theta)$, where the coefficients were connected by a three-term recurrence relation. His presentation of the analysis was somewhat involved, and certain simplifications were later made by Love. The work is, however, still quite complicated, and will not therefore be given here. Instead, we shall consider a special case in which the equation can be solved in finite terms (Solberg, 1936). This will illustrate the point that if the value of s is prescribed (for a given σ), then h must have one of a series of discrete eigenvalues. These were denoted in Chapter II by $h_r(s, \sigma)$, and the corresponding values of Θ by $\Theta_r^s(\theta)$.

We shall consider the axially symmetrical oscillation of period half a siderial day, i.e. we put $f=1$ and $s=0$ in equation (25) which becomes

$$\frac{1}{\sin\theta}\frac{d}{d\theta}\left(\frac{1}{\sin\theta}\frac{d\Theta_r^0}{d\theta}\right)+\frac{4a^2\omega^2}{gh}\Theta_r^0=0,$$

or, if $\mu=\cos\theta$,

$$\frac{d^2\Theta_r^0}{d\mu^2}+\frac{4a^2\omega^2}{gh}\Theta_r^0=0.$$

The solution of this equation is

$$\Theta_r^0=A\sin\left(\frac{2a\omega}{\sqrt{(gh)}}\mu\right)+B\cos\left(\frac{2a\omega}{\sqrt{(gh)}}\mu\right).$$

The condition imposed on the solution is that Θ_r^0 should be uniquely defined and single-valued, and have continuous derivatives. For this condition to be satisfied at the poles, $\mu=\pm 1$, it is necessary that

$$\frac{gh}{a^2\omega^2}=\frac{4}{n^2\pi},\tag{50}$$

where n is an integer. This result may be proved directly, but is more easily obtained by considering the case in which the ocean is confined by rigid boundaries to that part of the surface of the sphere lying between the circles of latitude given by $\mu=\pm\mu_1$. The condition to be expressed is then that the northward velocity u should be zero. Reference to equation (33) shows that since $s=0$, this requires $d\Theta_r^0/d\theta=0$, where $\mu=\pm\mu_1$. Hence it follows that

$$A\cos\frac{2a\omega}{\sqrt{(gh)}}\mu_1-B\sin\frac{2a\omega}{\sqrt{(gh)}}\mu_1=0,$$

$$A\cos\frac{2a\omega}{\sqrt{(gh)}}\mu_1+B\sin\frac{2a\omega}{\sqrt{(gh)}}\mu_1=0.$$

Thus either $\quad A=0 \quad$ and $\quad \sin\frac{2a\omega}{\sqrt{(gh)}}\mu_1=0,$

in which case the oscillation is symmetrical about the equator, or

$$B=0 \quad \text{and} \quad \cos\frac{2a\omega}{\sqrt{(gh)}}\mu_1=0,$$

when the oscillation is anti-symmetrical. In either case, we must have

$$\frac{2a\omega}{\sqrt{(gh)}}\mu_1=\frac{n\pi}{2},$$

where n is an integer. If μ_1 is now made equal to unity, so that the ocean covers the whole sphere, the result quoted above is obtained.

Hough gave the results of calculations made for four different values of $gh/4a^2\omega^2$, namely, $\frac{1}{40}$, $\frac{1}{20}$, $\frac{1}{10}$ and $\frac{1}{5}$, corresponding to values of h of 2·19, 4·38, 8·76 and 17·5 km. As explained above these results need correction before they can be applied to oscillations of the atmosphere, where mutual gravitation is negligible. Examination of the formulae, and a few trial calculations, shows that, except when $gh/4a^2\omega^2=\frac{1}{40}$ and $n=0$, the correction may be made with an accuracy of better than $\frac{1}{2}$% by multiplying the values of h for which the periods are given by Hough by $1-3\rho/(2n+1)\sigma$, where ρ and σ are respectively the densities of water and of the earth as a whole. Fig. 13 has been constructed in this way, except that a fresh calculation was made for one point in the top left-hand corner. A few independent calculations are to be found in the literature, and these, together with equation (50), have been used to provide a check.†

Two values of h of special importance are those corresponding to periods of 12 solar and 12 lunar hours respectively in the (2, 2) mode. These are

$$h=7{\cdot}9 \quad \text{(solar)}, \qquad h=7{\cdot}1 \quad \text{(lunar)}.$$

Fig. 13 gives most of the information likely to be required concerning the relation between h and σ. The position as regards the form of $\Theta_n^s(\theta)$ expressed as a series of harmonic functions is not, however, so satisfactory. Hough gave detailed results for $s=0$ for

† Slightly different values of a and g are taken by different writers; the following values are used here: $a=6{\cdot}37\times10^8$ cm., $g=980{\cdot}6$ cm./sec.^{2}, so that $4a^2\omega^2/g=$ 87·6 km.

the four depths he considered, but for other values of s contented himself with the remark that the general form was similar. As an

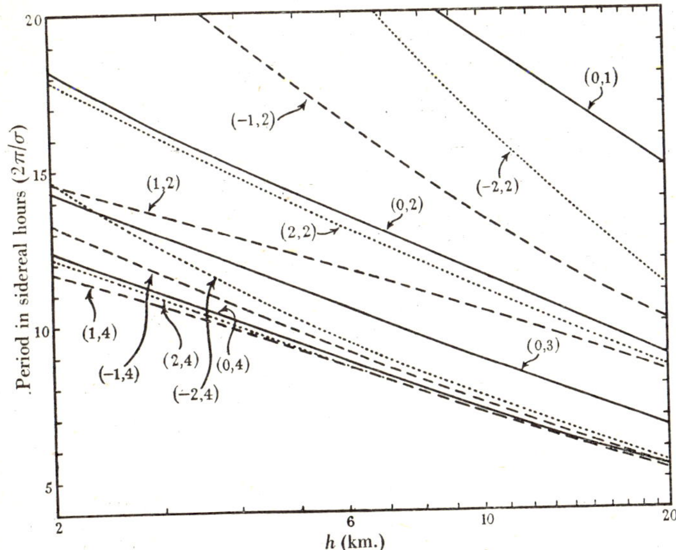

Fig. 13. The effective depth of the atmosphere $h = h_r(s, \sigma)$ as a function of period. The figures against the curves give the values of (s, r).

example, we may quote some of the actual results for $gh/4a^2\omega^2 = \frac{1}{10}$, $s = 0$:

$$\Theta_1^0(\theta) = P_1 - 0{\cdot}6516P_3 + 0{\cdot}1034P_5 - 0{\cdot}0073P_7 + 0{\cdot}0003P_9 - \ldots,$$

$$\Theta_2^0(\theta) = P_2 - 0{\cdot}4029P_4 + 0{\cdot}0477P_6 - 0{\cdot}0027P_8 + 0{\cdot}0001P_{10} - \ldots,$$

$$\Theta_3^0(\theta) = 0{\cdot}0471P_1 + P_3 - 0{\cdot}2726P_5 + 0{\cdot}0248P_7 - 0{\cdot}0011P_9 + \ldots,$$

$$\Theta_4^0(\theta) = 0{\cdot}0677P_2 + P_4 - 0{\cdot}1989P_6 + 0{\cdot}0144P_8 - 0{\cdot}0006P_{10} + \ldots,$$

$$\Theta_5^0(\theta) = 0{\cdot}0013P_1 + 0{\cdot}0689P_3 + P_5 - 0{\cdot}1547P_7 + 0{\cdot}0093P_9 - 0{\cdot}0003P_{11} + \ldots,$$

where P_n is written for $P_n(\cos\theta)$. These results have not been corrected for the effect of mutual gravitation.

It will be seen that the predominating term in $\Theta_r^0(\theta)$ is $P_r(\cos\theta)$. A similar result is true when s is not equal to zero; in this case the expansion proceeds in a series of associated Legendre functions $P_n^s(\cos\theta)$, and it is found that the term in $P_r^s(\cos\theta)$ predominates. This is illustrated by two results given by Pekeris (mutual gravitation excluded); the period is 12 solar hours in each case, the corresponding values of $gh/4a^2\omega^2$ being 0·0899 and 0·02420 respectively ($h=7{\cdot}87$ and 2·12 km.). The expressions have been normalized so that the coefficient of $P_r^s(\cos\theta)$ is unity:

$$\Theta_2^2(\theta)=P_2^2-0{\cdot}08697\,P_4^2+4{\cdot}821\times10^{-3}\,P_6^2-1{\cdot}656\times10^{-4}\,P_8^2 +3{\cdot}824\times10^{-6}\,P_{10}^2-6{\cdot}325\times10^{-8}P_{12}^2+\dots, \quad (51)$$

$$\Theta_4^2(\theta)=0{\cdot}7740\,P_2^2+P_4^2-0{\cdot}3747\,P_6^2+0{\cdot}06282\,P_8^2-6{\cdot}336\times10^{-3}\,P_{10}^2 +4{\cdot}339\times10^{-4}\,P_{12}^2-2{\cdot}165\times10^{-5}\,P_{14}^2+\dots, \quad (52)$$

where P_n^s stands for $P_n^s(\cos\theta)$.

It may be shown by a general argument that

$$\int_0^1 \Theta_n^s(\theta)\,\Theta_m^s(\theta)d(\cos\theta)=0 \quad (m\neq n).$$

If use is made of this result, the above formulae may be inverted to give expansions of $P_r^s(\cos\theta)$ in terms of Θ_r^s, Θ_{r+2}^s, etc. For example, we have

$$P_2^2(\cos\theta)=0{\cdot}9401\,\Theta_2^2(\theta)+0{\cdot}05972\,\Theta_4^2(\theta)+\dots \quad (53)$$

This holds only if the period of oscillation is equal to 12 solar hours; a similar result for 12 lunar hours cannot be given, but it is to be expected that the coefficients would be of a similar magnitude. Generally, it may be concluded that when h is large

$$\Theta_r^s(\theta)\sim P_r^s(\cos\theta).$$

This result is also true if h is not large, provided n is large.

The predominating component of the solar tide-producing potential is semi-diurnal, and has a distribution over the surface of the earth given by $1{\cdot}1\times10^4 e^{2i\phi}P_2^2(\cos\theta)$ in c.g.s. units. In view of (53) this may be written approximately $10^4\,e^{2i\phi}\Theta_2^2(\theta)$. In treating the solar semi-diurnal oscillation we may therefore put (see equation (26)).

$$\Omega(0)\simeq10^4$$

In the case of the moon the corresponding result is

$$\Omega(0)\simeq2\times10^4.$$

3·2. Evaluation of $y(x)$

Discussion of $\Theta_r^s(\theta)$ is to some extent simplified by the fact that the differential equation (25) from which it is derived does not involve the variation of temperature with height, so that solutions, once obtained, are applicable to any atmosphere. The corresponding equation for $y(x)$ (32) must, on the other hand, be solved afresh for each different atmosphere it is desired to consider.

In the general case of an arbitrary variation of temperature with height, equation (32) must be solved numerically or with the aid of a differential analyser. In certain special cases solutions may be obtained in terms of tabulated functions. For example, if H is proportional to x the solution for $y(x)$ involves Airy's integral and its associated function, while, if H varies linearly with height, the solution can be obtained in terms of a Bessel function. The latter form of solution was used by Pekeris to discuss the oscillations of an atmosphere in which the curve connecting temperature with height consisted of a series of straight lines.

In the case of a forced oscillation of the atmosphere a solution for $y(x)$ must be found which corresponds to an upward flow of energy at a level just below that at which the effects of absorption become important. It is convenient to assume that above this level the atmosphere is isothermal. This is justifiable since the actual conditions existing there are unimportant as far as the motion of the atmosphere lower down is concerned, provided there is no reflexion of energy. When the temperature at high level is low enough for $\frac{H}{h}\frac{\gamma-1}{\gamma}-\frac{1}{4}$ to be negative, the forced oscillation excited by a tide-producing force of given period may be evaluated by solving equation (32) subject to the condition that the solution tends to zero at infinity. If $\frac{H}{h}\frac{\gamma-1}{\gamma}-\frac{1}{4}=\mu^2>0$, the solution is complex, the real and imaginary parts satisfying the equation separately. In order to get the complete solution it is necessary to compute two numerical solutions, one approximating to $\cos(\mu x+\Phi)$ and one to $\sin(\mu x+\Phi)$ at infinity, and to take the first plus i times the second. If a differential analyser is used, this means that two runs of the machine must be made for each value of h under consideration.

When the complex solution for $y(x)$ has been obtained, the pressure variation at the ground, or at any other level, may be obtained from equation (36). A curve may then be plotted showing how the amplitude of the pressure oscillation varies with h, or with the period. Such a curve will be called a resonance curve for the atmosphere. It is convenient to express the amplitude of oscillation in terms of the 'equilibrium' tide, which is given by

$$\begin{aligned} p &= -\Omega\rho_g \\ &= -\Omega p_g/gH_g, \end{aligned}$$

where H_g, ρ_g and p_g denote the scale height, static density and static pressure at the ground.

The following simple expression for the pressure variation at ground level may be obtained from (36) and (38):

$$p_{z=0} = i\gamma p_g y(0)/\sigma.$$

CHAPTER IV

OSCILLATIONS IN THE ATMOSPHERE

4·1. The outward flux of energy

It has been shown in Chapter II that the tide-producing force can be resolved into a series of components, each of which excites a distinct mode of oscillation of the atmosphere. The resultant motion is then obtained by superposition. It may happen that one of the components is much larger than the others, or that it excites a resonant mode of the atmosphere. In either case the motion produced by this component predominates over that produced by the others, and it is therefore sufficient to consider it alone. In the following treatment it will be assumed that one mode only is excited.

A general appreciation in physical terms of the factors involved in atmospheric resonance may be obtained by considering the flux of tidal energy out of the atmosphere. Use has already been made of this method in discussing the nature of the boundary condition at high level. The greater part of this chapter will be devoted to the further application of this idea.

The energy put into the atmosphere by the tidal forces is nearly all introduced near the surface of the earth where the air density is high. This energy spreads out as a spherical wave in the atmosphere (see fig. 14). The motion is a complicated one, involving horizontal motion of the air particles as well as rarefaction and compression, but fortunately an examination of the equations developed in Chapter II shows that there is a simple analogy with the propagation of plane waves in a medium of varying refractive index.

It has been shown that if attention is confined to one mode of oscillation alone the pressure variation and velocity components may be expressed in terms of the hydrodynamical divergence $\chi(z, \theta, \phi)$ which separates into functions of z, θ, ϕ only. We wrote

$$\chi(z, \theta, \phi) = e^{is\phi}\Theta_r^s(\theta)\, e^{-\frac{1}{2}x}\, y(x),$$

where $x = \int dz/H$. $\Theta_r^s(\theta)$ gives the variation with latitude, and may be regarded as a known function. We are primarily concerned at present with the function $y(x)$ which it was shown in Chapter II satisfies the differential equation

$$\frac{d^2y}{dx^2}+\left[-\frac{1}{4}+\frac{1}{h}\left(\frac{dH}{dx}+\frac{\gamma-1}{\gamma}H\right)\right]y=0. \tag{54}$$

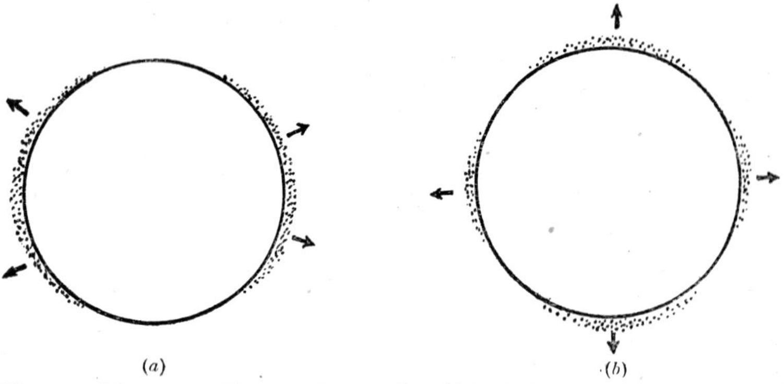

Fig. 14. Diagram to illustrate the way in which tidal energy introduced near the surface of the earth spreads out as a spherical wave. (*a*) and (*b*) refer to oscillations in which the exciting force varies with latitude in different ways.

It was further shown that, except near the surface of the earth, the rate of flow of energy in a vertically upward direction is given by

$$W \propto \mathscr{I} y^* \frac{dy}{dx}, \tag{55}$$

the constant of proportionality involving the function $\Theta_r^s(\theta)$.

These equations at once suggest an analogy with other forms of wave motion. For example, the propagation of electromagnetic waves in a medium having a refractive index

$$\frac{c}{\sigma}\left[-\frac{1}{4}+\frac{1}{h}\left(\frac{dH}{dx}+\frac{\gamma-1}{\gamma}H\right)\right],$$

where c is the velocity of light, depends on the solution of the differential equation

$$\frac{d^2E}{dx^2}+\left[-\frac{1}{4}+\frac{1}{h}\left(\frac{dH}{dx}+\frac{\gamma-1}{\gamma}H\right)\right]E=0, \tag{56}$$

where E is the electric vector, and the rate of flow of energy upward is given by

$$W \propto \mathscr{I} E \frac{dE^*}{dx}. \tag{57}$$

The coefficients in equation (56) are real, so that E^* satisfies the same equation, i.e.

$$\frac{d^2E^*}{dx^2}+\left[-\frac{1}{4}+\frac{1}{h}\left(\frac{dH}{dx}+\frac{\gamma-1}{\gamma}H\right)\right]E^*=0. \tag{58}$$

If (57) is now written

$$W\propto \mathscr{I}(E^*)^*\frac{dE^*}{dx}, \tag{59}$$

it will be seen that equations (58) and (59) are identical with equations (54) and (55), except that E^* appears instead of y. An exact mathematical analogy therefore exists between the two problems, and it follows that a solution of the problem in atmospheric oscillations may be obtained from one for the electric wave problem by taking y to be proportional to E^*. The rates of flow of energy in the two cases are then the same (subject to a multiplying constant), and we have a justification for applying physical arguments familiar in dealing with other types of wave motion to the propagation of tidal energy upwards in the atmosphere.

Similar considerations to those given above apply when the excitation is thermal instead of gravitational. Energy of oscillation is generated near the surface of the earth, and spreads upwards in the same way. An atmospheric oscillation can also be excited by the gravitational tides in the sea and in the body of the earth (Chapman, Pramanik and Topping, 1931). This may be allowed for by modifying equation (38) to allow for the motion of the earth's surface. It is convenient to refer to any oscillatory motion of the atmosphere as a tide, whether its origin is gravitational or not.

It is now possible, by considering the variation with height of the equivalent refractive index μ given by

$$\mu^2=\left[-\frac{1}{4}+\frac{1}{h}\left(\frac{dH}{dx}+\frac{\gamma-1}{\gamma}H\right)\right], \tag{60}$$

to see what are the conditions for resonance. Suppose that at some height μ^2 becomes negative and remains negative for all greater heights. Waves passing upwards are then totally reflected, and the energy introduced into the atmosphere by the tide-producing force is trapped, as though by a barrier. If the height of the barrier is suitably adjusted in relation to the period of the oscillation, the amplitude builds up, and resonance occurs. This may be otherwise expressed by saying that a free oscillation of this period exists.

If μ^2 is not negative for all heights above a certain level, but is negative only over a band of heights, the ordinary considerations of wave theory show that the barrier is partially transparent. A proportion of the oscillatory energy leaks through during each cycle and passes upwards. If there is no further barrier, it will ultimately enter a region of the atmosphere where the effects of viscosity and thermal conductivity become of importance, and will be absorbed. This energy is therefore lost to the part of the atmosphere whose oscillations we are considering, and these oscillations are consequently damped. If the barrier is only slightly transparent the resonance phenomenon will still be observed; if the barrier is almost wholly transparent, the resonance will be so damped as to be unobservable.

TABLE 7

T (° K.)	H (km.)	h (km.)	T (° K.)	H (km.)	h (km.)
180	5·26	6·01	260	7·61	8·75
200	5·86	6·70	280	8·20	9·38
220	6·44	7·35	300	8·79	10·05
240	7·03	8·05	320	9·37	10·70

Examination of equation (60) shows that μ^2 can become negative either on account of a low temperature, or a negative temperature gradient, or a combination of the two effects. In practice this implies a low-temperature region in the atmosphere. The situation is somewhat complicated by the fact that the effectiveness as a barrier of such a region depends markedly on the value of h, which in turn depends on the mode and period of the oscillation under consideration. As h is decreased, the barrier becomes rapidly more transparent, until for a certain critical value of h it ceases to be a barrier at all. Table 7 gives for reference purposes values of h above which μ^2 is negative in the case of an isothermal region. For any particular mode of oscillation, these values of h may be converted into periods by means of fig. 13.

Before discussing the oscillations of the actual atmosphere of the earth, it will be helpful to consider a hypothetical atmosphere consisting of a slab of air at temperature T_1 resting on another slab of air at temperature T_2 (see fig. 15).

If T_1 and T_2 are such that $\mu^2 < 0$ in the upper region and $\mu^2 > 0$ in the lower region, energy introduced near the ground will be trapped, and if the height of the discontinuity is suitably adjusted in relation to the period a free oscillation of the type shown at (*a*) will occur.

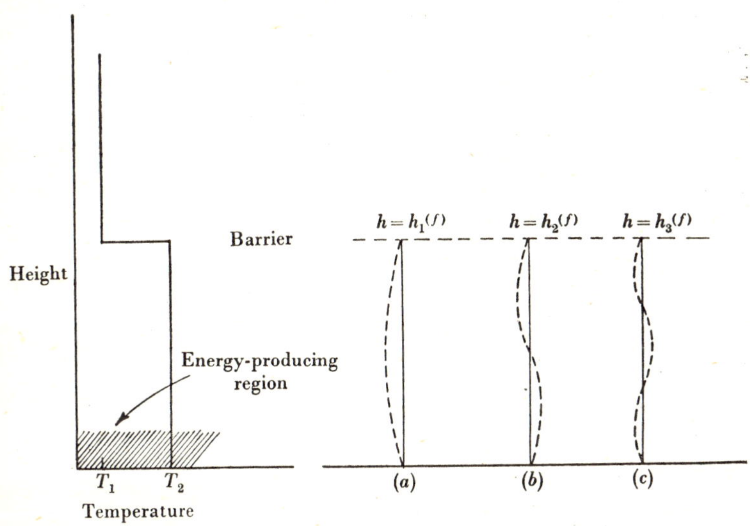

Fig. 15. Diagram to illustrate the trapping of energy in a simple atmosphere.

The period does not occur directly in this argument but is involved through the quantity h occurring in the expression for μ^2. The value of h corresponding to the free oscillation shown is an eigenvalue of equation (54), and in conformity with the notation used in Chapter II will be denoted by $h_1^{(f)}$. Since h depends on the type of oscillation as well as on the period, it follows that corresponding to $h_1^{(f)}$ there is a doubly infinite set of free oscillations with different periods given by fig. 13. All the oscillations have the same form as regards variation with height, but differ in their distribution with respect to latitude and longitude.

By analogy with the oscillations in an organ pipe it might now be thought that a second set of oscillations would exist of the type shown diagrammatically at (*b*) in fig. 15. It may easily be seen that these would correspond to a value of h (denoted by $h_2^{(f)}$)

which is smaller than $h_1^{(f)}$. Whether such a set of oscillations exists or not depends entirely on whether T_2 is sufficiently low for the condition $\mu^2 < 0$ to be satisfied for the new value of h. If it is not, oscillations of the type (*b*) will not exist, and there will only be one eigenvalue $h_1^{(f)}$. In a similar way oscillations with two, three or more nodes may exist corresponding to decreasing values of $h^{(f)}$; but sooner or later it will be found that the upper region ceases to be a barrier and the series of eigenvalues comes to an end.

4·2. The temperature variation in the atmosphere

Before we consider trapping of energy in the earth's atmosphere we will give a short account of the temperature variation with height as far as this is known.

The temperature in the lower atmosphere is known from direct observation with sounding balloons, which give information up to 16 km. as a routine matter, and on occasions, though with less certainty, up to 30 km. These data show that there is a uniform decrease of temperature from the surface of the earth up to the tropopause (which occurs at heights varying from 15 km. at the equator to 9 km. at the poles), after which the temperature is constant in temperate regions, and increases with height near the equator; at about 24 km. the temperature appears to be close to 220° K. over the whole earth, and above this height to be constant or very slowly increasing. Recent observations suggest that there may be appreciable daily variation of the temperature of the lower stratosphere amounting to as much as 5° K.

Until recently there have been no direct measurements of the temperature in the atmosphere above 30 km., and all knowledge of the variation above that level was obtained by indirect reasoning based on various observed phenomena. In the majority of these phenomena the quantity which is of importance is the local scale height $H = kT_0/mg$, where m is the mean molecular mass. This is also the quantity which occurs in equation (32) governing the oscillations of the atmosphere. If there is complete mixing, the scale height will be the same for all atmospheric constituents, but if there is any diffusive separation it will vary for different gases, and values deduced may not apply to the main mass of the atmosphere.

It is, however, unlikely that the diffusive separation is appreciable below 100 km., and we will assume that there is no diffusive separation.

The available information was summarized by Martyn and Pulley in 1936, and later in the report of a discussion on the upper atmosphere which took place at a joint meeting of the Chemical Society, the Physical Society, and the Royal Meteorological Society (*Quart. J. Roy. Met. Soc.* 1939). A more recent survey is given by Weekes and Wilkes (1947). Only a brief statement of the present state of knowledge will be given here.

The fact that sound waves are sometimes propagated to anomalously great distances can be explained if it is assumed that the temperature begins to rise again at about 35 km. (just above the highest level reached by sounding balloons), reaching a value of 300–350° K. at 50–60 km. This explanation is now generally accepted, although alternative suggestions have been made.

There is evidence based on the reflexion of very long wireless waves from the ionosphere (Budden, Ratcliffe and Wilkes, 1939; and Wilkes, 1940) that the temperature does not continue to increase above 60 km., but falls again, reaching a value as low as 200° K. at about 70 km., and that it is constant at this value up to 90 km. Evidence for a low temperature at about 80 km. has also been drawn from the existence of noctilucent clouds which are sometimes observed at this level (Humphreys, 1933; Regener, 1939–40). The observations are, however, rare, and there is uncertainty as to the composition of the clouds. It is often assumed that the temperature in the E region itself is about 350–400° K., corresponding to a scale height of 10–11 km., but the radio evidence on which this is based is not conclusive.

A further method by which information about atmospheric temperature may be obtained is by observation of auroral spectra. Difficulties of technique and interpretation have, however, so far prevented any reliable information being obtained by the method.

Very recently, measurements made in the high atmosphere by means of rockets have become available (Best, Havens and LaGow, 1947). Fig. 16(*a*) shows the pressure variation with height deduced from observations made during a flight on 7 March 1947, and fig. 16(*b*) the corresponding temperature curve. Pressures between

2 and 10^{-2} mm. were measured by means of two Pirani gauges mounted on opposite sides of the body of the rocket. A Phillips gauge, mounted on the cone of the warhead, measured the 'ram'

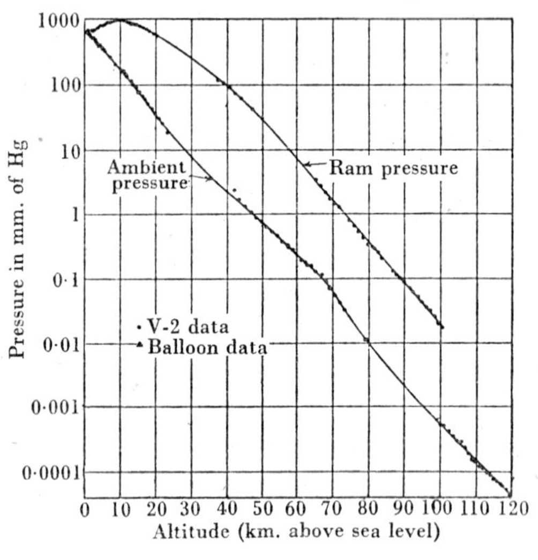

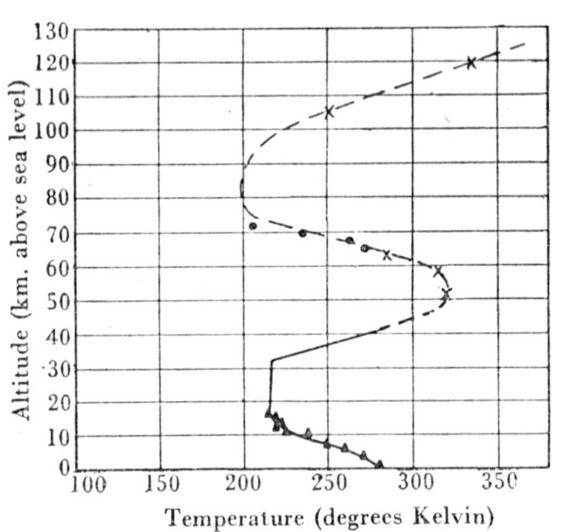

Fig. 16. Measurements made during a rocket flight on 7 March 1947 (Best, Havens and LaGow, 1947).

pressure from which ambient pressures in the range 10^{-3} to 10^{-5} mm. were derived by the use of Taylor and MacColl's theory. Temperatures were deduced both from the slope of the pressure curve, and from the ratio of the 'ram' pressure to the ambient pressure. It will be seen that the resulting temperature curve is precisely of the form which inference from the sound propagation and other data has led one to expect. This agreement is extremely satisfactory, especially when it is realized that the measurement of ambient pressure and temperatures from inside a rocket travelling at high speed through the highly rarified atmosphere presents a number of difficult technical problems—connected, for example, with time lags in the measuring instruments, and the effect of occluded gas escaping from the material of which the rocket is made.

Although the evidence outlined above gives a consistent picture of the temperature distribution in the atmosphere, there is difficulty in reconciling it with observations of meteor tracks. These would suggest that the temperature in the region 70–90 km. is somewhat higher—about 255°. There are, however, various theoretical difficulties in the interpretation of the observational results, and great weight cannot be attached to this evidence at the present time.

4·3. Oscillations of the earth's atmosphere (general treatment)

We may now consider trapping of tidal energy in the earth's atmosphere. It has already been explained that the greater part of the tidal energy is introduced near the surface of the ground; in practice this means the lower 15–20 km. of the atmosphere. Thus the troposphere and the lower part of the stratosphere constitute the energy-producing region. In the absence of any barrier this energy would pass upwards, and be finally absorbed at high level.

The temperature of the stratosphere is about 220° K.—corresponding to a scale height H equal to 6·44 km.—and reference to table 7 on p. 50 shows that it will constitute a barrier for oscillations corresponding to values of h greater than 7·35 km. There is thus a possibility that there will be a set of free oscillations formed by trapping of energy by the stratosphere. Numerical calculations, made using the known temperature variation in the troposphere and stratosphere, show that such a set of oscillations does in fact exist, the corresponding eigenvalue of h, which we will denote by $h_1^{(f)}$, being slightly greater than 10 km. It is to be noted that in this case the barrier extends right down to the region where the energy is introduced, there being no free region between. The calculations further show that no other type of free oscillation due to trapping by the stratosphere occurs; in fact, as h is reduced —i.e. for a given mode as the period is increased—the stratosphere rapidly becomes transparent.

If we now examine the periods of the various modes of oscillation corresponding to $h_1^{(f)}$ we find that there is no period nearly equal to 12 solar hours, as is required by the resonance theory. In fact the (2, 2) mode, which has a distribution over the earth's surface similar to that found for the solar semi-diurnal oscillation, has a period of approximately 10 hours. It thus follows that the requirement of the resonance theory cannot be met in this way.

It is to be noted that for values of h near 10 km. the stratosphere is a very good barrier, and very little energy leaks through. For this reason, the exact nature of the assumptions made as to the temperature variation above 30 km. is without appreciable effect on the value of $h_1^{(f)}$ obtained. Since the temperature variation in the atmosphere up to 30 km. is known by direct observation it follows that the above argument does not depend on any speculation as to the temperature in the atmosphere.

A check on the above theory may be obtained by considering the propagation of a pulse in the atmosphere. The wave produced by the Krakatoa eruption travelled with a velocity of 318·8 m./sec.; the velocity of the wave from the Great Siberian meteor was not determined with such great accuracy, but the mean value obtained was 318 m./sec. It was shown in Chapter II that the velocity of propagation, V, of a pulse is related to the eigenvalue $h^{(f)}$ by the relation $V^2 = gh^{(f)}$. If V is put equal to 318·8 m./sec., this gives $h^{(f)} = 10{\cdot}4$ km.; thus the existence of the eigenvalue $h_1^{(f)}$ with a value slightly greater than 10 km. is confirmed, though none of the free oscillations corresponding to it is excited to any appreciable extent in nature.

Returning to the resonance theory, we see that the only way in which the atmosphere can be made to have a free period in the (2, 2) mode of 12 solar hours is by assuming the existence of a second set of free periods corresponding to a second eigenvalue $h_2^{(f)}$. This in turn implies the existence at the higher levels of the atmosphere of a further barrier capable of trapping energy passing through the stratosphere.

It is already known from observations on the anomalous propagation of sound that the temperature rises again above the stratosphere. The existence of a second barrier therefore requires that the temperature must fall again at a still higher level. Pekeris showed by numerical calculation that there is no difficulty in adjusting such a 'cold top' to the atmosphere to give a value for $h_2^{(f)}$ of 7·9 km. corresponding to a free period of 12 solar hours for the (2, 2) mode. The oscillation is one which has a node in y at about 30 km., and it is found that the quantity $dy/dx - \frac{1}{2}y$ to which the pressure oscillation is proportional has also a node at a similar height. Trapping at the upper barrier cannot give rise to an oscillation with no node in y, since the relevant value of h would be large enough for very little energy to leak through the lower barrier. The details of these calculations are discussed in the next section.

We have seen in section 4·2 that there is independent evidence for concluding that a low-temperature region exists in the atmosphere round about 80 km. The above argument shows that the existence of such a region somewhere above the hot region inferred from the results on the anomalous propagation of sound, and below the level where viscous and other absorption becomes important, is an essential requirement for the validity of the resonance theory. It has been pointed out that this conclusion does not rest on any speculation as to the temperature variation in the atmosphere, and the existence of the low-temperature region must therefore be taken to be a direct consequence of the resonance theory.

It should be mentioned that the effect of the negative temperature gradient in the troposphere is to make μ^2 negative in this region for values of h as small as 4·2 km. although the stratosphere itself is transparent if h is greater than 7·4 km. This fact has been ignored in the descriptive argument given above on the grounds that the region is not very thick, and that much of the tidal energy is produced in or even above it. Its effect is, of course, fully taken into account in the numerical work which has been quoted.

One point remains to be dealt with. It was shown in Chapter II that a pulse can be propagated without change of form with velocity

$\sqrt{(gh^{(f)})}$, and it follows that if there are two values of $h^{(f)}$ there will be two possible velocities for the propagation of a pulse. We should therefore expect the Krakatoa explosion to have given rise to two separate waves travelling with velocities 319 and 280 m./sec. respectively, and it is necessary to explain why the records show only one wave, corresponding to $h_1^{(f)} = 10{\cdot}4$ km.

Pekeris (1939) considered this problem, and pointed out that the bulk of the energy in the faster pulse has its seat lower down than in the case of the slower pulse. This is to be expected from the general argument given above, since trapping of energy takes place in the stratosphere in the first case, and around 80 km. in the second. If the explosion takes place near the surface of the earth, the faster pulse should therefore be excited more strongly than the slower one. Pekeris considered the matter quantitatively, and found that the ratio in the case of the Krakatoa eruption should be about 5:2. He examined the Krakatoa records for evidence of the second wave at the expected level and found that, although its existence could not be definitely asserted, the evidence was by no means inconsistent with its existence.

Pekeris estimated that the total energy of the atmospheric wave excited by the Krakatoa eruption was about 10^{24} ergs. It is interesting to compare this with the energy of the atomic bomb explosion at Nagasaki which was stated at the time to be equivalent to the explosion of 20,000 tons of T.N.T. or 10^{21} ergs. It is not known how much of this energy passed into a travelling air-wave of the type considered here.

4·4. Oscillations of the earth's atmosphere (numerical results)

We now pass to more detailed consideration of the temperature variation in the atmosphere in relation to the resonance theory, and for this purpose we shall make use of calculations made in the Mathematical Laboratory, Cambridge, by means of a differential analyser (Weekes and Wilkes, 1947).

A convenient starting-point is to take the atmosphere given by Martyn and Pulley in 1936. This was based on a general survey of the data available at that time, and was revised slightly by Martyn in 1939 (see fig. 17). Calculation shows that, as would be expected on the general grounds of the last section, this atmosphere has two eigenvalues $h^{(f)}$, namely, $h_1^{(f)} = 10{\cdot}2$ km. and $h_2^{(f)} = 8{\cdot}8$ km. respectively. The first is in sufficient agreement with the value 10·4 km. deduced from the velocity of the Krakatoa wave, but the second is well removed from the value of 7·9 km. required by the resonance theory. It thus appears that Martyn and Pulley's atmosphere is not consistent with the resonance theory.

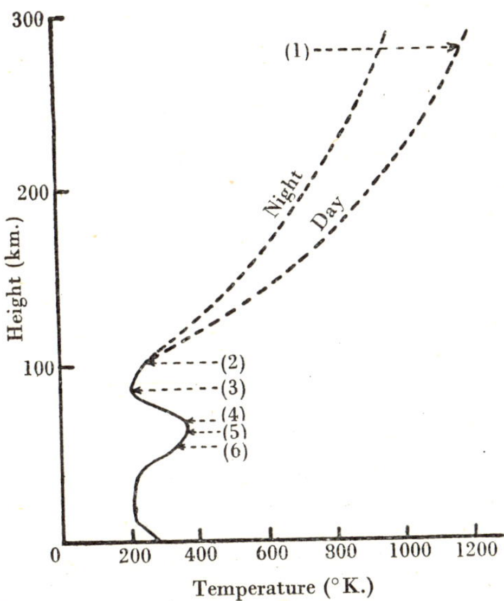

Fig. 17. Temperature variation in the atmosphere proposed by Martyn (1939). (1) Electron collision frequencies (Martyn and Pulley), F_2 thickness (Fuchs, Appleton). (2) Electron collision frequencies (Baily and Martyn). (3) Noctilucent clouds (Humphreys). Reflexion of very long radio waves (Budden, Ratcliffe and Wilkes). (4) Atmospheric oscillations (Pekeris). (5) Meteors (Lindeman and Dobson). (6) Abnormal sound wave propagation (F. J. W. Whipple).

In considering what modifications are necessary we shall confine ourselves to atmospheres in which the curve connecting temperature with height consists of a series of straight lines. In this way the number of parameters needed to specify a given atmosphere is reduced to a minimum, and in view of the limited amount of information at present available any more elaborate representation is scarcely justified. Needless to say the differential analyser enables solutions to be obtained for any temperature distribution, and various tests were made to make sure that the use of segmented distributions gave rise to no peculiar effects, such, for example, as might arise from the existence of discontinuities in the temperature gradient.

We begin by considering atmospheres of the type shown as *ABCDEFG* in fig. 18, where the temperature at high level is sufficiently low for μ^2 to be negative when h is greater than or equal to 6·4 km. The temperature variation up to 30 km. was

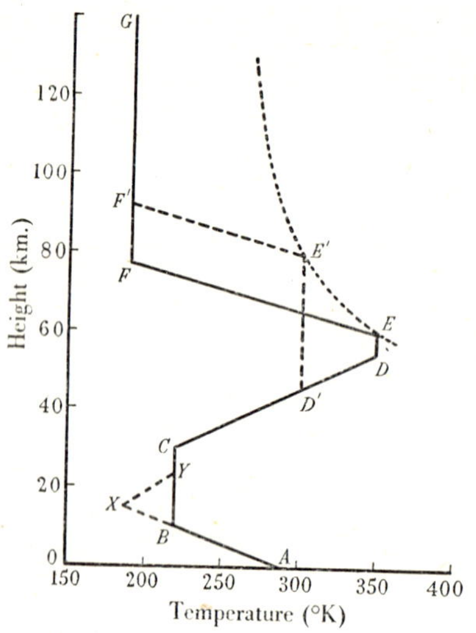

Fig. 18. The simplest type of atmosphere which has a free period of type (2, 2) with a period of 12 solar hours.

chosen so as to agree with observation, and the various 'tops' which would make the atmosphere as a whole have a value of $h_2^{(f)}$ equal to 7·9 km. studied. The value of $h_1^{(f)}$ is mainly determined by conditions in the atmosphere very low down, and is little affected when the temperature above 30 km. is changed.

In one series of calculations, of which the results are summarized in fig. 18, the temperature gradients above and below the maximum at about 60 km. were kept constant while the actual value of the maximum and its thickness were altered. It was found that any atmosphere whose temperature variation is given by a curve like *ABCDEFG*, where the point *E* lies on the dotted curve, satisfies the requirements of the resonance theory in that it has free oscillations with values of $h^{(f)}$ equal to 7·9 and 10·4 km.

In fig. 18 the temperature gradients above and below the maximum are taken to be $9\frac{2}{3}$ and $5\frac{1}{2}$°K./km. respectively. If, instead, the rate of rise is taken to be 13° K./km. starting from 35 km. instead of 30 km., the resulting dotted curve is found to be nearly parallel to that shown in fig. 18, but moved down 4 km. in height. A higher temperature for the isothermal top reduces the width of the temperature maximum by about 1 km. for 10° K. change in temperature. The temperature of the isothermal portion must, of course, be less than 240° K. in order to ensure that there shall be a barrier.

The calculations described were made for an atmosphere in which the temperature variation in the stratosphere was similar to that observed in temperate latitudes. Over a considerable part of the earth, however (including the tropics, where the tide-producing force has its maximum value), the temperature variation is rather different (see section 4·2). The calculations were therefore repeated for a series of stratospheres in which the temperature variation was given by curves like *ABXYCDEFG*. It was found that in each case the effect on the free oscillation corresponding to $h_2^{(f)} = 7{\cdot}9$ km. was very small, and could be compensated for by a slight alteration to the temperature at the 60 km. level; the actual change was of the order of 0·2%. The effect on the oscillation corresponding to $h_1^{(f)} = 10{\cdot}4$ km. was rather larger, being equivalent to a 1% change in $h_1^{(f)}$; this, however, is within the basic accuracy of the theory.

Calculations have also shown that alterations in the temperature of the stratosphere by ±10° K. have only a small effect on the 7·9 km. mode, and do not alter the form of the vertical distribution of pressure variation in the oscillation to an appreciable extent. The case in which the stratosphere temperature changes with latitude and longitude cannot be treated by present methods, but the results recorded in this section suggest that the effect of allowing for this variation would be small.

The atmospheres so far considered are bounded above by a perfect barrier, so that the amplitude of the forced oscillation increases without limit as the resonant frequency is approached. It has been pointed out in section 4·1 of this chapter that if the temperature, having fallen to a low value, rises again higher up, the barrier will be partially transparent, and energy will leak through, to be absorbed at higher level still where damping becomes important.

The point is illustrated numerically in fig 19, which shows resonance curves calculated for a number of atmospheres differing only in the height at which the rise begins to take place. It will be

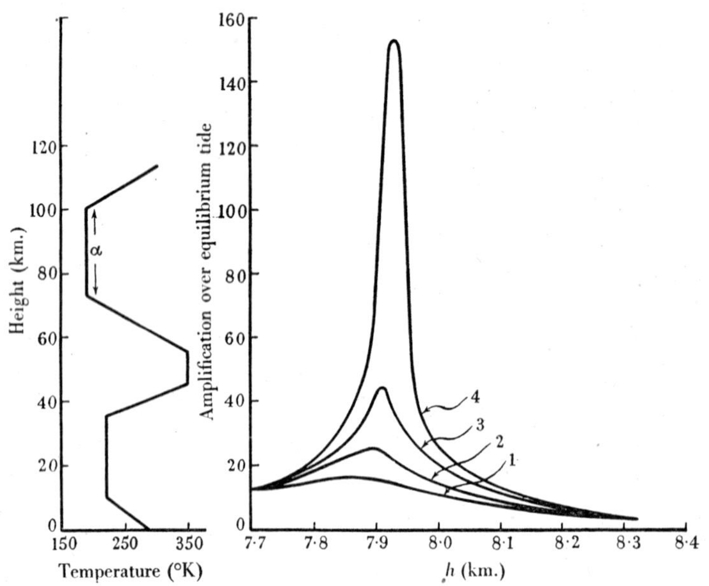

Fig. 19. Diagram showing the effect of variation of barrier thickness on sharpness of resonance. Curve 1, $a=13$ km.; curve 2, $a=18$ km.; curve 3, $a=23$ km.; curve 4, $a=28$ km.

seen that the higher this is, i.e. the thicker the barrier, the sharper is the resonance. As with other resonant systems, the skirts of the resonance curve are affected to a comparatively small extent by the amount of damping present.

The action of the barrier is due partly to the region of falling temperature above 53 km. and partly to the isothermal region; in the case of the widest barriers, these contributions are approximately equal. A rough measure of the effectiveness of a barrier is given by the value of the quantity $\int \mu dz$ taken over the barrier region; the results indicate that for an amplification of 70 this must have a value of about 2·2.

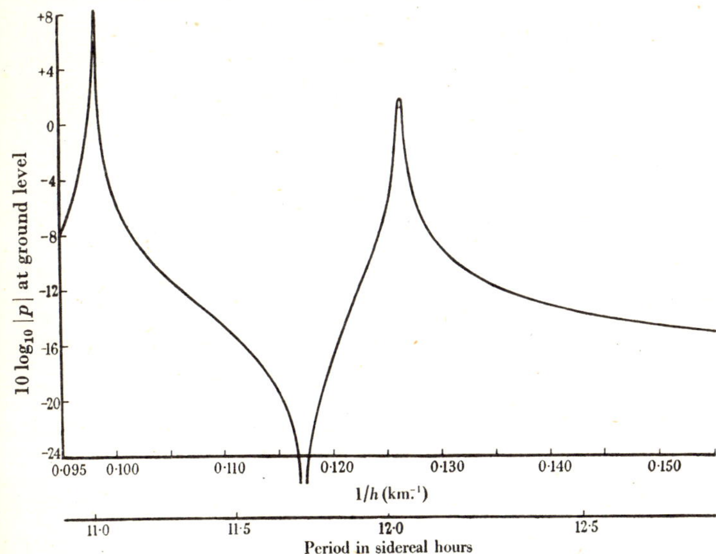

Fig. 20. Resonance curve for atmosphere 4 of fig. 3 plotted over a wider range of equivalent depth h. The pressure variation is measured in mm. and $\Omega(0)$ is taken to be 10^4 c.g.s. units.

TABLE 8

z (km.)	T (° K.)	H (km.)	x	e^{-x}	p_0 (mm.)
0	288	8·4	0	1	760
10	220	6·4	1·4	0·25	190
20	220	6·4	2·9	0·055	42
30	220	6·4	4·5	0·011	8·4
40	272	8·0	5·9	0·0027	2·0
50	324	9·5	7·0	$9·1 \times 10^{-4}$	0·69
60	350	10·2	8·0	$3·3 \times 10^{-4}$	0·25
70	256	7·5	9·2	$1·0 \times 10^{-4}$	0·076
80	190	5·6	10·8	$2·0 \times 10^{-5}$	0·015
90	190	5·6	12·5	$3·7 \times 10^{-6}$	$2·8 \times 10^{-3}$
100	190	5·6	14·3	$6·2 \times 10^{-7}$	$4·7 \times 10^{-4}$
110	230	6·7	15·9	$1·2 \times 10^{-7}$	$9·1 \times 10^{-5}$
120	230	6·7	17·4	$2·8 \times 10^{-8}$	$2·1 \times 10^{-5}$

Fig. 20 shows the resonance curve for the deepest barrier plotted over a wider range of h. The scale of period in hours is for the (2, 2) mode, and the ordinate scale gives the value of $10 \log_{10} |p|$

at the ground for this mode of oscillation, where p is measured in mm. $\Omega(0)$ is taken as 10^4 c.g.s. units, which corresponds to the solar semi-diurnal gravitational potential (see section 3·1).

For convenience, data are given in table 8 for an atmosphere which has correct resonance properties and which agrees as well as may be with observational indications as to temperature variation, including those from rocket experiments. The atmosphere is the one shown as *ABCDEFG* in fig. 18, except that the temperature begins to increase again at 100 km., reaching a value of 230° K. at 110 km.

CHAPTER V

DISCUSSION OF RESULTS

It is intended in this final chapter to discuss the extent to which the theory of air-tides developed in the preceding chapters is sufficient to explain the experimental facts. Broadly, it may be said that the general features of both the lunar and solar tides are well explained, but that the theory throws little light on such aspects as seasonal and annual changes. There are also some difficulties connected with the air-tide at very high level, i.e. in the ionosphere.

The resonance theory may now be taken as well established. The great regularity, symmetry, and large amplitude of the solar semi-diurnal barometric variation have always been strong arguments in its favour, and the results of rocket experiments and other evidence have now shown that the temperature of the atmosphere varies with height in the right kind of way to give a free period of 12 solar hours. The velocity of propagation of very long air-waves is also given correctly by the theory, and this provides a check on certain of the fundamental assumptions.

A difficulty in the way of the resonance theory when it was first put forward was to see how the necessary accuracy of tuning could be maintained in spite of the large variations in the temperature and pressure of the atmosphere which are associated with changes in the weather (F. J. W. Whipple, 1918). It is now realized that these changes are very localized in nature, and affect only the first 15 km. or so of the atmosphere, whereas the solar semi-diurnal oscillation is world-wide in extent and embraces the atmosphere up to the height of the E region. The vast scale of the oscillation also provides the answer to another question put by Whipple, namely, how it is possible for the wave to surmount the heights of Central Asia and the Rocky Mountains at each revolution, without losing an appreciable proportion of its energy.

It is not yet possible to say to what extent the solar semi-diurnal tide is due to gravitational and to what extent to thermal action of the sun. If gravitation were the sole cause, the amplification over the equilibrium tide required would be about 100, and reference to fig. 19 shows that this would be quite feasible. However, Chapman (1924) has shown by a simple argument depending on the observed phase lead of the tide on the passage of the sun, that the thermal contribution must be at least of the same order of magnitude as the gravitational one. The fact that thermal action is important reinforces the argument in favour of the resonance theory based on the observed uniformity of the oscillation, since the thermal exciting force itself must vary greatly over the earth's surface, particularly from land to sea.

In two important respects there are difficulties in reconciling the theory as so far given with observation. An account was given in section 1·4 of the determination by Appleton and Weekes of the lunar semi-diurnal variation in the height of the E region (104–118 km.). It was found to have an amplitude of 0·93 km., and to be approximately in phase with the barometric tide at Greenwich, high pressure corresponding to a high observed height. The large amplitude, which seems surprising at first, can be explained by the theory, but there are difficulties about the phase.

It is usually assumed when discussing the formation of the E region that the ions produced by absorption of solar radiation have a very short life. It follows that a given ionization density, if the solar zenith angle is assumed constant, will always occur at a level where the pressure has a certain fixed value. If the level at which this pressure occurs moves up and down as a result of tidal motion, so will the region as a whole. If the observed variation of height of the region from the mean is a, the corresponding pressure variation at a fixed height is $-ap_0/H$. The results of Appleton and Weekes, interpreted on this basis, lead to the conclusion that the pressure variation in the E region is in phase with that at the ground, and not out of phase as would be expected from oscillation theory.

If it is assumed that the ions have a long life so that the tidal effect makes itself felt through the physical transport of ions from one level to another, it is possible to find a solution to the difficulty. The vertical velocity w given by equation (35) of Chapter II does not change sign with height as do u and v, and it may be shown that the motion of the air particles is in the right direction to account for the phase of the lunar tide in the E region, and that the movement is of the right order of magnitude. Unfortunately, it is difficult to see how, on any acceptable theory of the ionosphere, the life of the ions can be long enough for this explanation to be correct. Martyn (1947, 1948) has, however, pointed out that an additional vertical component of velocity will be produced by the horizontal motion of the ions across the horizontal component of the earth's magnetic field. This may well be sufficiently great to produce a displacement of the ionised layer of the observed order of magnitude in spite of the rapid rate of recombination.

The experiments of Appleton and Weekes also established the fact that at the level of the E region the ratio of the amplitudes of the solar and lunar oscillations is less than at the ground. Consideration of the solar and lunar quiet day magnetic variations leads to a similar conclusion. These variations, according to the dynamo theory first propounded by Balfour Stewart, are due to the motion of ionized air under tidal action across the earth's magnetic field. It is found that the two variations are in many ways similar, but that in certain respects—notably dependence on season and sun-spot cycle—their behaviour is somewhat different (Chapman and Bartels, 1940, Chaps. VII and VIII). This suggests that different parts of the ionosphere contribute differently to the lunar and solar variations, which in turn implies that the lunar and solar air-tides are in different ratios at different heights. It must, however, be mentioned that recent work has suggested that the difference in behaviour of the lunar and solar variation is not so great as had been formerly believed (Bartels and Johnston, 1940).

The phases of the magnetic variations are such as would be expected if the lunar and solar tides at the levels concerned were out of phase with those at the ground. These results, therefore, unlike those of Appleton and Weekes, do not conflict with oscillation theory, and in fact provide strong evidence in its support.

The work of Martyn on the lunar tide in the E region has already been mentioned, and it appears probable that his explanation of why the phase of the tide in the E region is opposite to what would be expected according to oscillation theory is the correct one. It is, however, worth while showing that, even if Martyn's theory is not accepted, and if the results of Appleton and Weekes are taken to indicate that the pressure variation in the E region is in phase with that at the ground, it is still possible to reconcile theory with experiment. The magnitude of the pressure variation is then equal to about 10% of the static pressure, the exact value depending on the value taken for H. The relative pressure variation p/p_0 is thus about 10^4 times as large as at the ground.

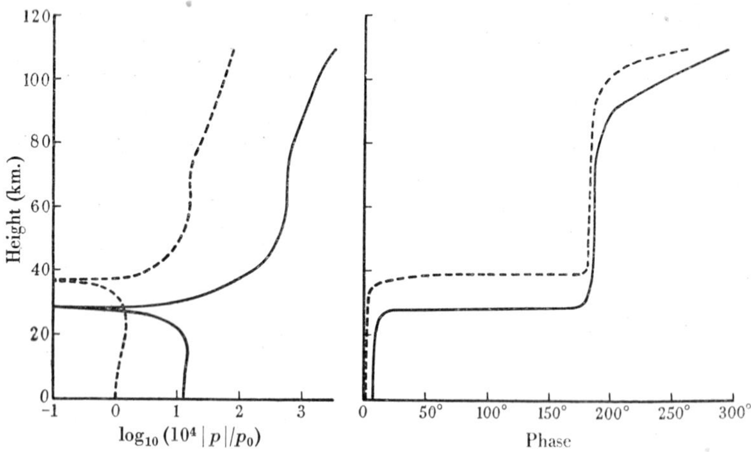

Fig. 21. The amplitude and phase of the pressure variation in atmosphere 4 of fig. 19 plotted as a function of height. For oscillations of type (2, 2) the dotted and solid curves refer to periods equal to the lunar and solar half days respectively.

In the first place, it is clear from fig. 21 that in atmospheres of the type considered in Chapter IV, the solar and lunar oscillations are in much the same ratio at all heights, provided it is assumed in each case that the (2, 2) mode is excited to much greater extent than the other modes. This assumption would appear to be justified—in the case of the sun by the resonance theory, and in the case of the moon by the fact that the predominating term in the expansion of the tide-producing potential is such as would excite this mode. If, however, other modes were excited to an appreciable extent in the lunar case, the dependence of the pressure oscillation on height might be modified considerably, since each mode would correspond to a different value of h and therefore to a different function $y(x)$. Until further calculations have been made this possibility cannot be entirely ruled out, although at the present time it seems unlikely.

Apart from the consideration just mentioned, however, it is to be noted that the calculations quoted so far are based on the assumption that the temperature in the E region is independent of height. If, instead, a hot E region with a falling temperature above is assumed, quite different results are obtained. Any assumption about the temperature in the E region is, of course, highly specula-

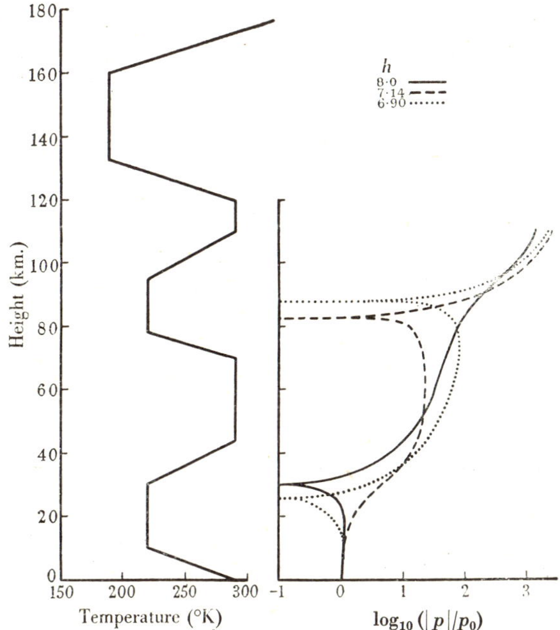

Fig. 22. Pressure variation as a function of height in an atmosphere having a temperature maximum in the E region for three different values of equivalent depth h. The curves are adjusted to agree at the ground.

tive, but there is nothing unreasonable in assuming that the temperature rises to a maximum in the E region, in view of the large amount of solar energy being absorbed in that region.

The effect of having a falling temperature above a hot E region is to give rise to a further barrier which may, if suitably adjusted, trap energy which penetrates the two lower barriers. It introduces the possibility of a third set of free periods corresponding to a third eigenvalue, $h_3^{(f)}$. The results of some calculations made for such an E region are shown in figs. 22 and 23. It will be seen that the solar and lunar tide-producing forces excite oscillations (of the (2, 2) mode) having different vertical distributions of pressure variation, that for the sun having one pressure node as before and that for the moon having two such nodes. The solar oscillation at high level will then still be out of phase with that at the ground, while that

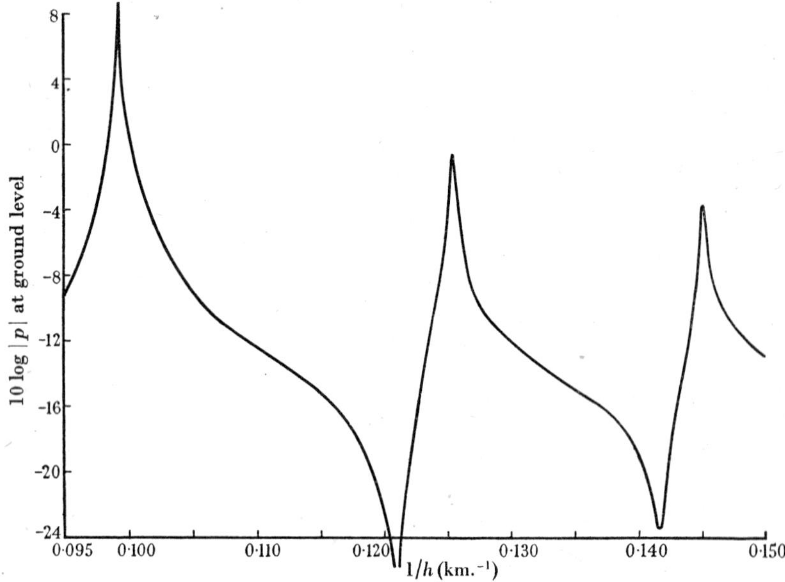

Fig. 23. Resonance curve for the atmosphere shown in fig. 22. p is measured in mm. and $\Omega(0)$ is taken equal to 10^4 c.g.s. units.

for the moon will be in phase. Curves for the variation of pressure with height are given in fig. 22. It will be seen that the curves for the solar and lunar cases are very different, and the differences in the solar and lunar quiet day magnetic variations can therefore be explained.

The atmosphere shown in fig. 22 was chosen in order to show that there is no difficulty in reconciling the experimental result of Appleton and Weekes with theory and must not be taken as representing a firm suggestion for the temperature variation in the E region. Calculations for other atmospheres of a similar type have yielded results of the same kind.

The subject of tidal motion in the E region is of great interest, and measurements similar to those of Appleton and Weekes, but made in different latitudes, would be of great value.

Martyn has shown by analysis of observational data that tidal effects can be found in the F_1 and F_2 regions of the ionosphere. Theoretical treatment of tidal motion in these high regions involves two considerations which have been ignored in the theoretical treatment given in Chapter II. One is that since p/p_0 increases with height the assumption that the oscillation is small is not true above a certain height. The other is that motion of ionized air across the earth's magnetic field gives rise to electrodynamic forces. Viscous and conductivity damping are of great importance at these high levels on account of the low air density.

The amount of energy required to excite and sustain an oscillation in the F region is probably very small on account of the low density. How much of this energy is provided by leakage from the lower atmosphere through the barrier below the E region, and how much is directly introduced by thermal or gravitational action are matters for further consideration.

We now pass to discuss the variations of the solar barometric oscillation over the surface of the earth. It will be remembered from Chapter I that Schmidt and Simpson analysed the oscillation into two components—a travelling wave and a wave stationary with respect to the earth. The first of these waves may be identified with an oscillation of the resonating (2, 2) mode, and its amplitude would be expected to vary with latitude like $\Theta_2^2(\theta)$ calculated for $h=7{\cdot}9$ km. (equation (51), Chapter III).

Table 9 shows a comparison of the theoretical values with values calculated from the empirical formulae given by Schmidt and Simpson (equations (1) and (3) of Chapter I respectively). In each case the figures have been multiplied by a suitable scaling factor to make them agree at the equator.

TABLE 9

Latitude $90°-\theta$	Theoretical	Schmidt	Simpson
90°	0·0000	0·000	0·000
80°	0·0005	0·013	0·005
60°	0·040	0·142	0·125
40°	0·27	0·45	0·45
20°	0·73	0·82	0·83
0°	1·00	1·00	1·00

The figures calculated from the formulae of Schmidt and Simpson are in general agreement, and show that the amplitude of the oscillation falls off much less rapidly with increasing latitude than theory would suggest. It must be remembered that the theoretical formula was obtained on the simplifying assumption that the temperature variation in the atmosphere was independent of latitude. It is possible that the error so introduced may be sufficient to account for the discrepancies revealed in table 9.

Little progress can be made in considering theoretically the stationary component of the barometric oscillation. It may be thought of as made up of the sum of a number of oscillations of type (0, r), and in view of the existence of surface features on the earth it is quite understandable that such oscillations should be excited.

Examination of fig. 13 shows that, corresponding to $h_2^{(f)}=7{\cdot}9$ km., there are several modes—apart from the (2, 2) mode—which have free periods equal to simple submultiples of 12 solar hours, e.g. the (2, 4) mode for which the period is 8 hours.† There is thus the possibility that other modes, besides the (2, 2) mode, are resonant, and several writers have examined the observational data for evidence of this.

The subject is, however, complicated by the fact that too much reliance cannot be placed on the exact shape of the curves of fig. 13 in view of the neglect of the fact that atmospheric temperature varies with latitude as well as with height. On the observational side, while a case may be made out for the occurrence of resonance in one or two cases (e.g. the 8-hourly component, although it does not have the form (2, 4)) the evidence is not nearly as convincing as in the case of the (2, 2) semi-diurnal mode.

† It has been suggested that one of the (0, r) modes resonates at 12 solar hours, thus accounting for the stationary solar oscillation, but this is not in fact the case.

REFERENCES

AIRY (1877). *Greenwich Meteorological Reductions*, 1854–73, *Barometer*. London.

APPLETON and WEEKES (1939). *Proc. Roy. Soc.* A, **171**, 171.

BARTELS and JOHNSTON (1940). *Terr. Magn.* **45**, 269; 485.

BERGSMA (1871). *Batavia Observations*, **6**, App. 2.

BEST, HAVENS and LAGOW (1947). *Phys. Rev.* **71**, 915.

BEZE (1690). *Mem. Acad. Sci.* **7**, 839.

BRUNT (1939). *Physical and Dynamical Meteorology*. Camb. Univ. Press.

BUDDEN, RATCLIFFE and WILKES (1939). *Proc. Roy. Soc.* A, **171**, 188.

CHAPMAN (1918). *Quart. J. Roy. Met. Soc.* **44**, 271.

CHAPMAN (1919). *Quart. J. Roy. Met. Soc.* **45**, 113.

CHAPMAN (1924). *Quart. J. Roy. Met. Soc.* **50**, 165.

CHAPMAN (1939). *International Union of Geodesy and Geophysics, Association of Meteorology*. Presidential address.

CHAPMAN and BARTELS (1940). *Geomagnetism*. Oxford.

CHAPMAN and TSCHU (1948). *Proc. Roy. Soc.* A, **195**, 310.

CHAPMAN, PRAMANIK and TOPPING (1931). *Beitr. Geophys.* **33**, 246.

ELLIOT (1852). *Philos. Trans.*, 125.

HANN (1889). *Denkschr. Akad. Wiss. Wien*, **55**, 49.

HANN (1892). *Denkschr. Akad. Wiss. Wien*, **59**, 297.

HANN (1918). *Denkschr. Akad. Wiss. Wien*, **95**, 1.

HOUGH (1897). *Philos. Trans.* A, **189**, 201.

HOUGH (1898). *Philos. Trans.* A, **191**, 139.

HUMPHREYS (1933). *Mon. Weath. Rev.* **61**, 228.

KELVIN (1882). *Proc. Roy. Soc. Edinb.* **11**, 396.

LAMB (1910). *Proc. Roy. Soc.* A, **84**, 551.

LAMB (1932). *Hydrodynamics*. Camb. Univ. Press.

LAPLACE (1823). *Mécanique céleste*, **5**, 149. Paris.

LOVE (1913). *Proc. Lond. Math. Soc.* (2), **12**, 309.

MARTYN (1939). *Quart. J. Roy. Met. Soc.* **65**, 328.

MARTYN (1947). *Proc. Roy. Soc.* A, **189**, 241.

MARTYN (1947). *Proc. Roy. Soc.* A, **190**, 273.

MARTYN (1948). *Proc. Roy. Soc.* A, **194**, 429; 445.

MARTYN and PULLEY (1936). *Proc. Roy. Soc.* A, **154**, 455.

PEKERIS (1937). *Proc. Roy. Soc.* A, **158**, 650.

PEKERIS (1939). *Proc. Roy. Soc.* A, **171**, 434.

PRAMANIK (1926). *Mem. Roy. Met. Soc.* **1**, 35.

RAYLEIGH (1890). *Phil. Mag.* (4), **29**, 173.

REGENER (1939–40). *Jb. dtsch. Akad. Luftforsch.* 262.

SABINE (1847). *Philos. Trans.*, 45.

Schmidt (1890). *Met. Z.* **7**, 183.
Schmidt (1919). *Met. Z.* **36**, 29.
Simpson (1918). *Quart. J. Roy. Met. Soc.* **44**, 1.
Solberg (1936). *Astrophys. Norwegica,* **1**, 237.
Taylor (1936). *Proc. Roy. Soc.* A, **156**, 318.
Weekes and Wilkes (1947). *Proc. Roy. Soc.* A, **192**, 80.
Whipple, F. J. W. (1918). *Quart. J. Roy. Met. Soc.* **44**, 20.
Wilkes (1940). *Proc. Roy. Soc.* A, **175**, 143.

INDEX